How to choose a plastics injection moulding machine
by Alan Smith

How to choose a plastics injection moulding machine
by Alan Smith

AMI Business Publishing

PROFESSIONAL PRACTICE SERIES

Series editor- John Goff

How to choose a plastics injection moulding machine

First Published 1995 in Great Britain by
AMI Business Publishing,
a division of
APPLIED MARKET INFORMATION LTD.
AMI House,
45-47 Stokes Croft,
Bristol BS1 3QP..

ISBN 1 871463 38 6
Printed in Great Britain by BPC Wheatons Ltd, Exeter

contents

foreword

For a quick briefing on the essentials

The busy reader may study the illustrations on pages 11-26, then read Chapter 1 and, if possible, Chapters 2, 3 and 4 as well.

But if you have the time -

Other readers may first study the illustrations but next read Chapter 5 before Chapter 1. The remainder of the book can be read in the order as presented.

The order of the contents of this book was carefully considered but there was a problem in reconciling the needs of readers with limited time available and others able to read it through in its entirety at relative leisure.

All readers, assuming that they start with little or no knowledge of injection moulding machines, are recommended first to study pages 11-26 on which annotated diagrams illustrate the main features of a typical machine. This should make them familiar with the layout of the machine.

Chapter 1 - a step by step guide to machine selection - immediately follows these illustrations for the benefit of readers with little time to spare and who want a quick briefing on how to extract the essential information from a manufacturer's specification table. The busy reader is also strongly recommended to find time to read Chapters 2, 3 and 4 if possible. Chapter 4, in particular, deals with the importance of screw performance, a vital aspect.

Other readers may prefer to read Chapter 5 after studying the illustrations and before reading Chapter 1. Chapter 5 describes the various designs of machine on the market and should enable the contents of Chapter 1 to be more readily understood.

introduction

This book sets out to help anyone who wants to buy an injection moulding machine but does not know where to start. It is not addressed to technically trained people but beginners, many of whom discover the injection moulding process for themselves and decide to use it to manufacture a novel product. Other readers may be those who find themselves suddenly plunged into the world of injection moulding and need a crash course in the subject. The author found himself in this position some years ago.

The first chapter goes directly into a description of how to extract the essential details from a machinery manufacturer's table of specifications, but two preliminary points need to be made. One is that the prospective machine buyer need not do this at all. He can simply go to a machinery supplier, say what he wants to make and leave it to him to recommend a suitable machine.

The second point is that, other than describing how to make a simple initial production cost analysis, this book does not go deeply into deciding upon the most suitable size of mould, a fundamental question which needs to be answered before buying a machine of any kind. This matter is dealt with in various books (see 'Further reading' page175) and there are also a number of specific computer software programs to assist. Major suppliers of machinery have access to production simulation software and can help potential customers in this respect, so it is unnecessary to go into a subject that is adequately covered elsewhere. The following text assumes that the prospective buyer has already decided upon the size of his mould or moulds and their cavitation (the number of cavities), the type of material he is going to use and is now seeking suitable machinery.

A buyer will want to obtain the best possible machine for his money, but it is important to keep in mind that the performance of a machine depends primarily upon the quality of the mould or moulds run on it. This question is addressed in Chapter 6.

There is no such thing as a best machine in the absolute sense, although there will be a machine, or class of machine, which is within the buyer's means and is best suited to the manufacture of a particular product. As this book is intended to be an introductory primer to assist the newcomer in selecting an injection moulding machine, only those machine features which need to be considered at this stage are discussed and a number of other aspects of injection moulding are not dealt with. The 'Further reading' section lists several authoritative works to which the reader in search of more detailed knowledge of the process may refer.

Where to see machines.

Trade directories which give the names and addresses of machinery suppliers are given in the 'Further reading' section. Machines can be seen in the showrooms of suppliers, most of whom are in the centre of England and only a few minutes off a motorway. A large supplier will normally have several machines on show to represent his range and some organise periodic open days with special demonstrations. There are also opportunities for prospective buyers to take part in group visits to the plants of Continental manufacturers, particularly German, Austrian and Swiss, when the companies concerned demonstrate and present lectures on their latest technology.

In most years there is at least one major plastics exhibition in Europe. Kunststoff, known simply as K, is the chief international event and is held every three years, including 1995, at Dusseldorf. On a smaller scale but still enabling a wide range of machines to be seen, the UK national exhibition, Interplas, is held in the next year after K at the National Exhibition Centre, Birmingham. Other important shows are Plast Milan, Europlast Paris and the National Plastics Exhibition in Chicago.

▲ About the author

Alan Smith was born in Acton, Middlesex, in 1929 and went to Isleworth County Grammar School from 1940 to 1945. He began his career in technical journalism with Temple Press, London, served in Greece with the Royal Army Educational Corps and later undertook public relations consultancy for various industrial concerns. These included Peco Machinery, a onetime major British manufacturer of injection moulding machines and extruders. His work for this company, which began in the early 1960s, was his introduction to the plastics industry, and in the mid 1970s he was co-founder and joint managing director of Butler-Smith Associates, makers of injection moulding machines. He subsequently became sales agent to Albis UK for its range of ancillary equipment. In the 10 years preceding his retirement in 1994, he returned to journalism and as machinery editor of *Plastics & Rubber Weekly* published many descriptions of the activities of plastics processors and machinery manufacturers in Europe, North America and the Far East. He was also responsible for the widely read 'Machine on test' series of reports published between 1985-89. These involved conducting comparative assessment trials of injection moulding machines of various makes.

A Fellow of the Royal Society of Arts, he served for some years on the Processing and Engineering Committee of the Plastics and Rubber Institute (later incorporated in the Institute of Materials). He is a contributor to a number of plastics and other technical journals.

colour section

The annotated diagrams in this section may be studied as a preliminary to reading the main text.

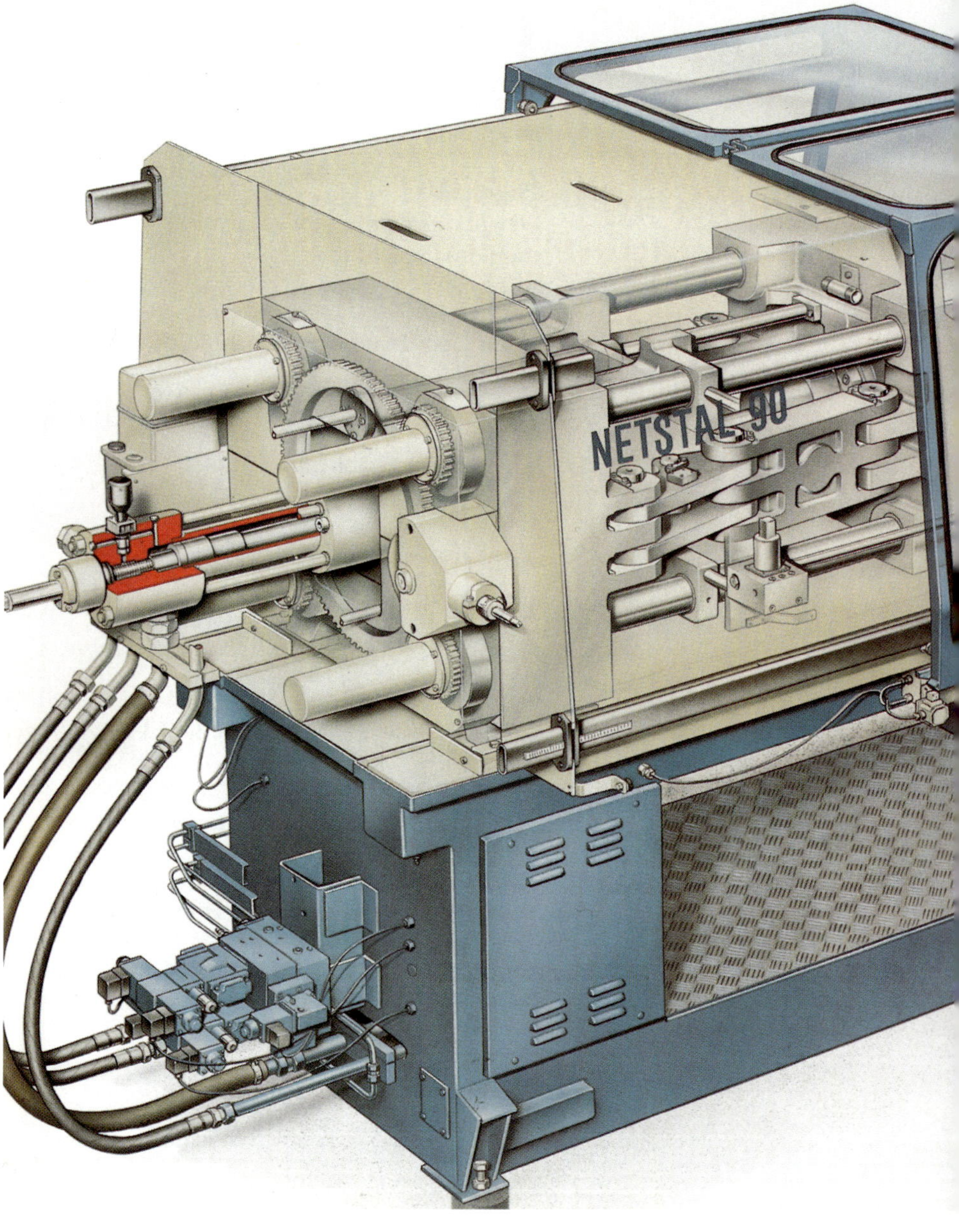
NETSTAL 90

NETSTAL
NETSTAL
SYCAP

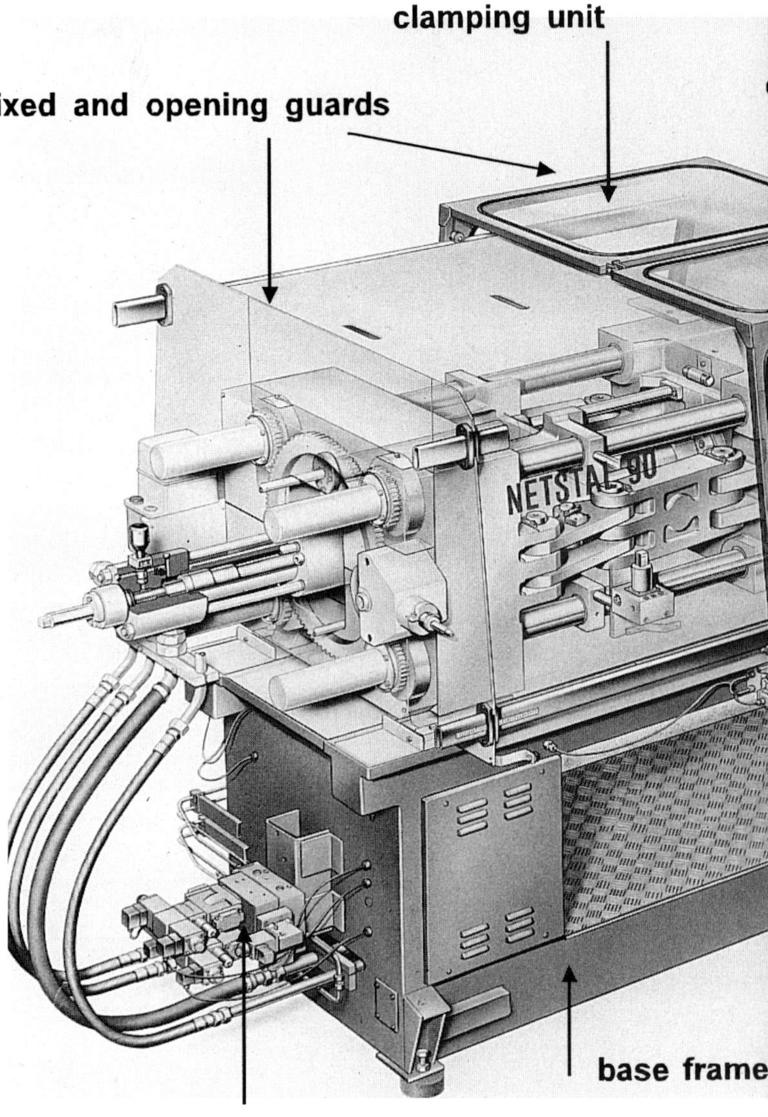
clamping unit
fixed and opening guards
NETSTAL 90
base frame
hydraulic valve block for clamping unit

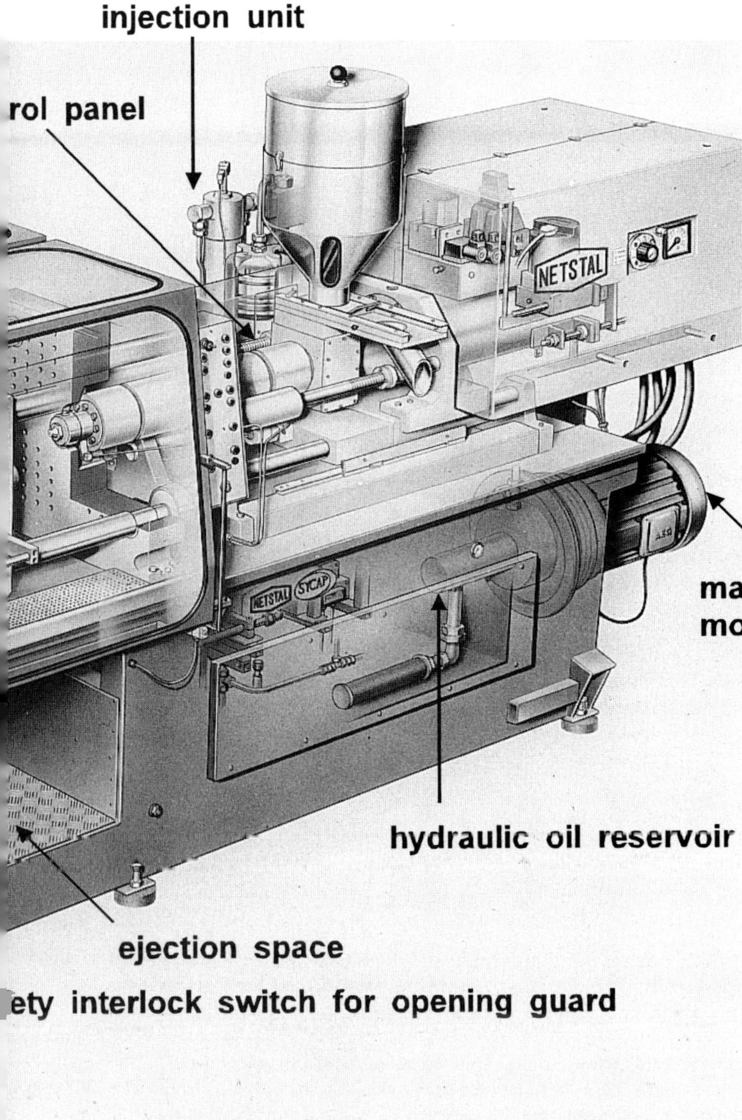

The clamping and injection unit assemblies are mounted on a base frame and there are fixed and opening guards over moving parts that may be a danger to personnel. The safety interlock switch ensures that the machine cannot be run unless the opening guard is closed.

The mould is mounted in the clamping unit and the space beneath this is called the ejection space. A box or belt conveyor for the collection of moulded parts may be placed in here.

The hydraulic oil reservoir is beneath the injection unit and there is a removable plate on the front for access to the pump and main filter. The control panel depicted is a small console on the machine and conveniently located for setting, that is adjusting the machine to the requirements of the mould. There is also a separate main control box with a VDU screen, not shown in this illustration. Other machines may have an integral main control box on the front.

Injection moulding machines are conventionally oriented with the clamping unit on the left and the injection unit on the right. This view therefore shows the front of the machine, that is the long side with the control panel.

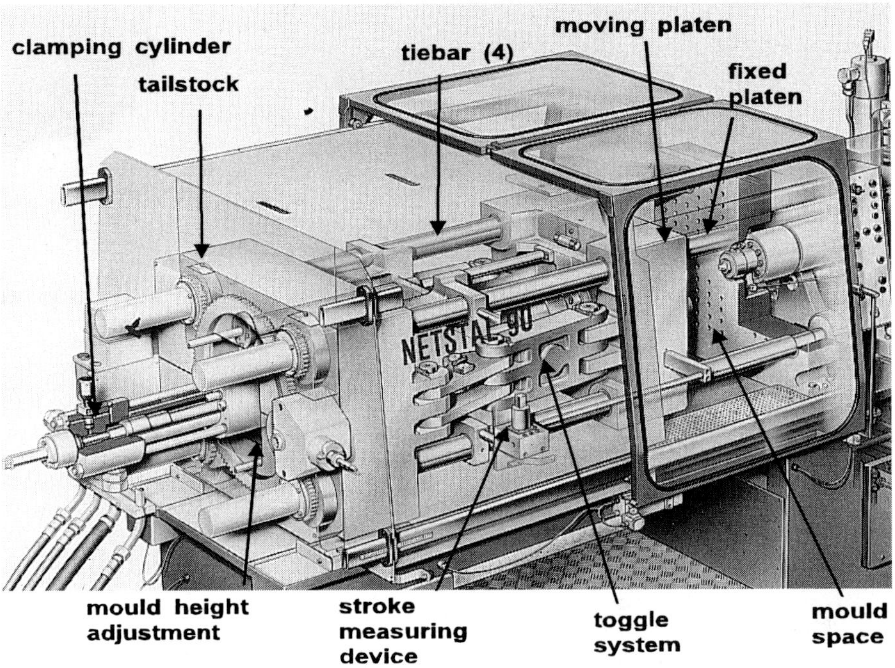

Clamping unit.
The toggle clamping system is fully extended in this drawing and the moving platen is seen moved forward to close the mould (not shown). The clamping cylinder and piston actuate the toggle system, which works like a lazy tongs to push the moving platen forward and subsequently pull it back to open the mould.
The moving platens slides backwards and forwards on the four tiebars (also known as 'columns'). The fixed half of the mould is mounted on the fixed platen and the moving half on the moving platen. The mould is thus closed and clamped tightly together for the injection of material at high pressure. It is then opened so that the moulded part or parts can be ejected. The distance through which the mould travels for closing and opening the mould is called the stroke and is precisely controlled by means of the stroke measuring device. Mould height adjustment is explained in the main text.
There are threaded bolt holes in the platens for mounting the halves of the mould. The fixed platen is cut away in this drawing in order to show the end of the injection unit but some of the bolt holes can be seen. The space between the platen faces is called the mould space.

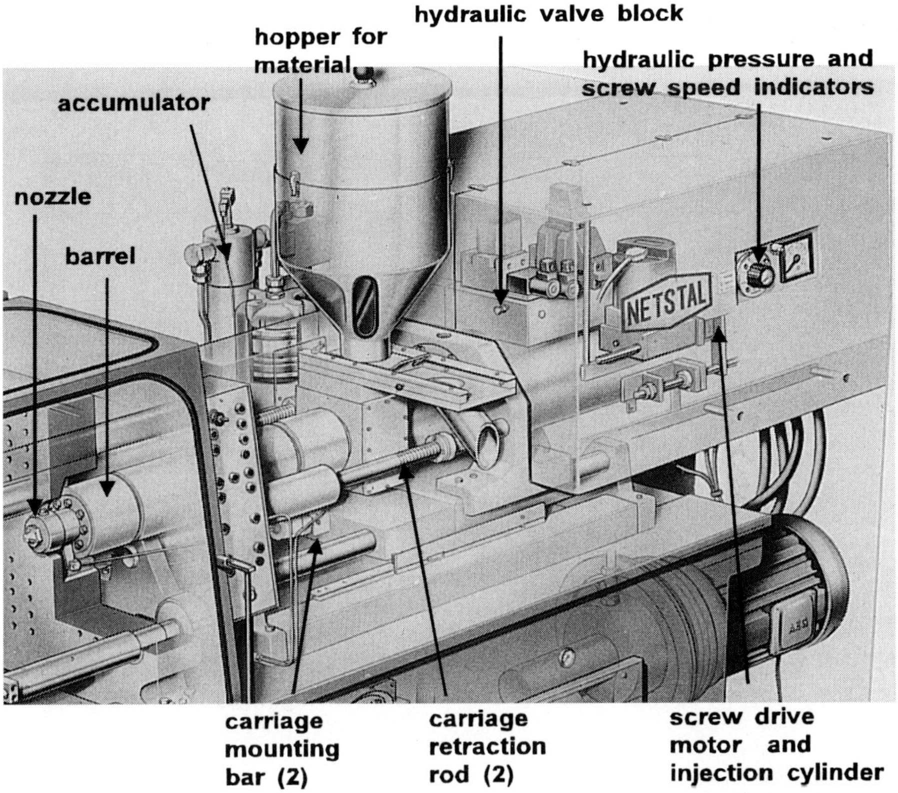

Injection unit.

The barrel of the injection unit is a heavy cylinder with a central hole or bore in which the plasticising screw rotates. This action puts heat into the plastics granules, drawn from the hopper, and melts them. A desired amount of molten material is then injected into the mould, with the screw acting as a kind of syringe or plunger. The nozzle at the front protrudes through a round hole in the fixed platen so that it can be placed directly against the sprue bush, or short entry tube, into the mould. Electric heater bands around the barrel and the nozzle maintain temperature when the screw is not rotating.

The injection cylinder and the motor that rotates the screw are obscured in this drawing but their location is indicated.

The machine here has an accumulator to boost hydraulic pressure.

The whole of the injection unit is mounted on a pair of support bars and can be moved towards and away from the fixed platen. The injection unit is referred to as the 'carriage' and a pair of hydraulic cylinders and rods provides for this movement. The carriage is advanced to bring the nozzle into contact with the sprue bush for injection and it may be retracted for a short distance according to the method of moulding being used. Further retraction allows access to the front end of the barrel for servicing. An American term for the carriage is the 'sled'.

The hopper for material can be moved towards the front of the machine so that any material left in it can be emptied out through the angled discharge tube seen. When installed in a moulding shop, a machine will usually have an automatic hopper loader operated by vacuum to keep the hopper topped up with material.

Hydro mechanical machine.

Hydro mechanical machines are of various designs but all have a two stage clamping system. On this 13,000 kN machine the moving platen is brought forward by a pair of relatively small hydraulic rams and then clamped to the four tiebars. Two of the tiebars can be seen here and are the heavy columns at the top and bottom of the picture. The moving platen is the massive piece at the centre. To the left it can be seen that the tiebars are deeply threaded behind the moving platen like coarse screws. These threaded sections enable pairs of jaws to close around them and secure the platen to the tiebars.

The first stage of clamping is completed in this way and the second stage is achieved when four cylinders on the injection unit side of the fixed platen pull the tiebars and the moving platen attached to them forward for a short distance to close the mould. These cylinders cannot be seen in the picture as they are on the other side of the fixed platen, the large flat plate on the right with a central hole. A further point of interest is that the platen has a series of T-slots for mounting the mould instead of rows of threaded bolt holes. Mounting bolts can be entered into these slots and positioned where required.

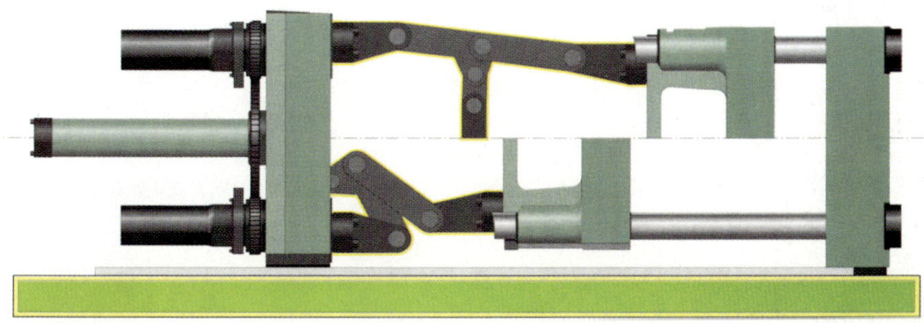

Toggle system.
This diagram, in which the tiebars are partly cut away for clarity, shows how a five point toggle clamping system works. In the upper half of the drawing the toggle links are fully extended and the moving platen is pushed forward to its farthest extent. In the lower the links have been brought back and folded and the moving platen is at its rearmost limit. The distance over which it has travelled is called the clamping stroke.

Hydraulic injection unit swing out.
The injection unit of many machines is pivoted so that it may conveniently be swung out to the front so that the nozzle or the screw shut off valve can be serviced. Hydraulic power for swinging out the unit is either standard or offered as an option

Retractable tiebar.
A large mould may be fitted in a machine by removal of an upper tiebar, or both upper tiebars, but this is a lengthy job. Some machines therefore have means for automatic tiebar retraction to save time. As shown in this picture, the front upper tiebar of the machine has been drawn back from the mould space. This is an optional feature for large machines.

Double daylight machines.
Machines with two mould spaces allow simultaneous production of paired components and for maximum use to be made of the injection unit. Two moulds are mounted in the machine and open and close alternately. The injection unit is at the back of the machine and injects material into each mould by means of a hot runner system built into the central platen. This has valves to direct the melt alternately into each mould.

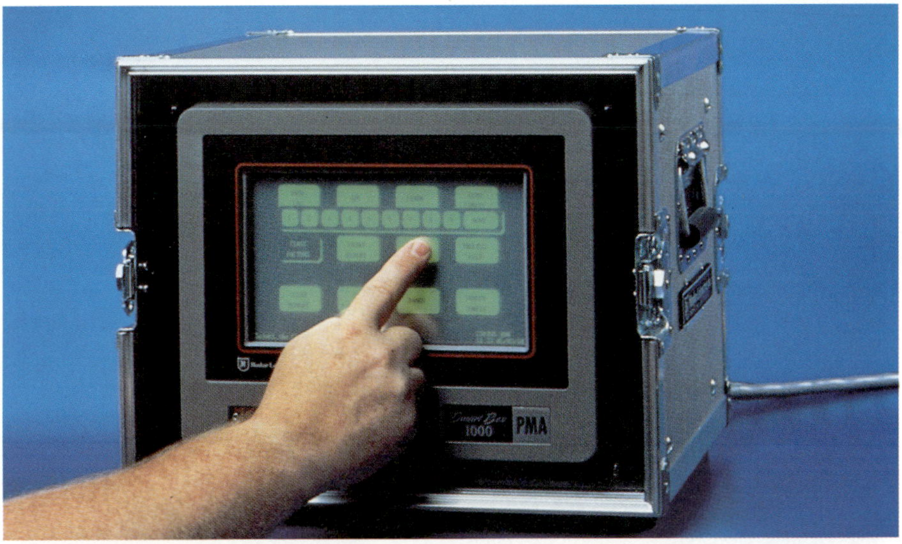

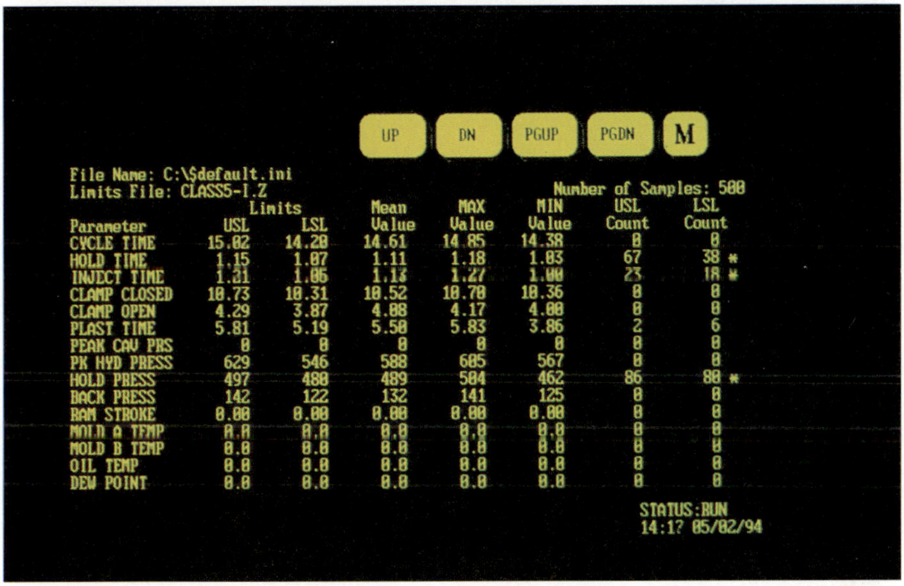

A portable machine analyser enables the performance of different machines to be compared.
Screen display on a portable machine analyser for a machine test run over 500 cycles. 'USL' and 'LSL' are the upper and lower specification limits. The results show how often these were exceeded or not attained, for example injection time exceeded the USL on 23 cycles and was below the LSL for 18 cycles. New machines of the same make and model can vary in performance, moreover performance of a new machine can deteriorate after only a few months running so that retesting within the guarantee period for a machine is advised. A drift away from original performance can mean slower cycling and lower profitability. It is also possible to use this equipment to match moulds to the most suitable machines to make best use of available moulding shop plant. Moulds demanding high performance are run on top quality machines and less demanding moulds on machines with lower performance.

▲ Mouldings and machine sizes

These illustrations provide a guide to the size or machine in terms of kN clamping force required for various typical mouldings. Cycle times quoted indicate the capability of the injection moulding process.

1	2
3	4
5	6
7	8

1.Tamperproof caps are produced in moulds with multiple cavities as high volume production is necessary to meet mass demand. This cap is 45 mm in diameter and the mould in which it is made has 16 cavities, that is 16 caps are produced for each cycle of the machine. The projected area for one cap has to be multiplied by 16 in order to calculate the total projected area. The 900 kN machine used provides both adequate clamping force and sufficient platen area for the mould. Cycle time is 6 seconds.

2. These aerosol caps are 50 mm in diameter and made in a mould with 10 cavities run on a 1,400 kN machine, also on a 6 second cycle.

3. A 1,400 kN machine is also used for these beakers but they are greater in diameter than the aerosol caps so that there are fewer cavities in the mould. The diameter is 75 mm and the mould has four cavities. Cycle time is 4 seconds.

4. This handle for an electrical isolator is 170 mm long and the round end is 70 mm in diameter. The mould has two cavities and a 1,400 kN machine is used. Cycle time is extended because of the substantial wall thickness. This calls for a fairly long cooling time.

5. Parts for industrial applications, as distinct from consumer markets, are called 'technical' or 'engineering' mouldings. Many of these are components that were previously made in metal and this pump housing is an example of a conversion to plastics. It measures 140 mm long by 110 mm wide and is made in a single cavity mould on a 1,400 kN machine. This has more than sufficient clamping force relative to the projected area of the moulding but large platens are needed for the core pulling attachments to the mould which are used to form the holes in the moulding that are not in the line of draw, that is the direction in which the mould is opened.
The core pulling attachments place the cores in position before material is injected and withdraw them once the moulding has cooled. The mould can safely be opened only once the cores have been withdrawn. Substantial wall thickness again extends cycle time, as applies to many technical mouldings.

6. Measuring 320 mm in diameter, this bucket is made in a single cavity mould on a 3,000 kN machine. Cycle time is 18 seconds.

7. A 3,000 kN machine is also used to make this instrument panel but the mould for it has two cavities. Each panel is 340 mm long and 140 mm wide so that the projected area for the two is about the same as for the bucket moulded singly on a machine of similar size. This moulding also features thinwall ribbing to reinforce it. Cycle time is extended because dimensional accuracy to 0.01 mm is required and a long cooling time is necessary to ensure this.

8. This mushroom box is made in a mould with special inbuilt sliding mechanisms on each of its four sides and required a machine with a large platen area relative to clamping force. The sides of the mould have rows of square blocks, corresponding to the hole patterns formed, but these need to be withdrawn to allow the box to be demoulded. Part of each side of the mould is therefore made as a moveable piece that is pulled outward and away from the centre by the special mechanisms, for which sufficient platen area has to be provided. The moulding is 360 mm long by 275 mm wide and a 4,500 kN machine is used. Cycle time is 8.5 seconds.

list of illustrations in main text

chapter 1

A step by step guide to choosing a machine. The most important headings in a machine specification table.

Choosing an injection moulding machine starts in respect of the clamping unit and figures for the following should be written down:-
1. Clamping force required in kN. (How to work this out is explained below.)
2. Mould width in mm.
3. Mould length in mm.
4. Mould height or thickness in mm.
5. Opening stroke required in mm.
There are two further points, 6 and 7, which concern the injection unit and will be dealt with later.

▲ A few preliminary words about terminology

We are dealing here with ***injection*** moulds, so described to distinguish them from moulds employed in other plastics conversion processes such as blow moulding and sheet thermoforming. A mould may also be referred to as a 'tool', and mouldmakers sometimes call an injection mould a ***mould*** tool to distinguish it from a ***press*** tool as used in metal forming.

'Die' is occasionally used to mean an injection mould but to avoid confusion this word is better confined to extrusion technology. In this book only 'mould' or 'tool' is used.

If the machine is intended for a number of moulds of different sizes, this list can be completed to cover all of them. It is possible to interchange screws and barrels with different shot volume capacities on the same machine. A machine may be obtained with two or more screw/barrel assemblies to suit several moulds and the most suitable assembly is fitted interchangeably. The moulds should all be of about the same size for the same machine.

It is also possible to standardise upon the sizes of what are known as the backplates for the moulds. These are the outer support plates for the tool itself and are bolted or clamped to the platens. One may have a number of moulds for a range of various products but the backplates are all of similar dimensions and so fit in the one size of machine.

1. Clamping force.

We first have to define how big a machine we want. The larger the moulding that one wishes to make, the bigger the machine and its clamping force required. The size of the moulding is defined by its weight and what is called its projected area, a term which will be explained in the following paragraphs. Weight will be dealt with when we come to point 6.

'Projected area' is best explained by example. An object such as a round beaker is moulded in a machine in a horizontal attitude, that is at right angles to its normal upright position and with its rim towards the moving platen. If it was possible to place one's eye in the centre of the moving platen, the view of the beaker that one would see would be of its projected area. In this case it would be a simple circle and the size of the area could easily be calculated.

A slimmer round object such as the barrel of a ballpoint pen would have a much smaller projected area. 'Project' means to throw forward, so that the projected area of a moulding is the area that it throws forward in the direction of the moving platen according to the attitude in which it is moulded. The projected area of a moulding of complex shape can, at the most basic level, be measured by placing the outline on graph paper and counting the squares covered, but as most mould design is done by CAD this figure can be worked out as part of the process.

Without going too deeply into the technicalities involved, it may be explained that the projected area of a moulding affects how much pressure is put against the moving platen when molten plastics material (known as the 'melt') is injected into the mould. Broadly, this pressure is irrespective of the weight and amount of material injected. Reverting to our examples, a beaker would put much more pressure against the moving platen during injection than the pen barrel because its projected area is many times greater. A force has to be applied to resist this pressure, which is known as the hydrostatic pressure of the melt.

Melt hydrostatic pressure has to be resisted if the two halves of the mould are to be held together securely in the clamping unit and the melt prevented from leaking out, a phenomenon called 'flashing'. The clamping unit is a kind of large vice with means for being closed and opened quickly. 'Halves' is not an exact term in the sense that each half is equal in size to the other but simply means the two main parts of the mould, one fixed and the other moving.

To make an accurate calculation of clamping force required, we should also take what is known as 'cavity pressure' into account, but as discussion of this at the present stage would make matters too complicated the whole matter is separately described at length in the Appendix on page 171.

There was a rough and ready rule for calculating clamping force in the early days of injection moulding and for the moment we will take this. The projected area of the moulding in square inches was simply multiplied by 4 tons. For example, if a moulding had a projected area of 20 sq in, the clamping force required for the machine was reckoned to be 80 tons. Today we would express this rule metrically as 6.2 kN (kilonewtons) of clamping force per cm^2 of projected area. The kilonewton is the recognised unit of force.

Injection moulding machines are made in various sizes designated by their maximum available clamping forces. If, for example, one needs a clamping force of 1,156 kN for a particular moulding, a machine with at least this amount must be selected, say 1,350 or 1,500 kN, either of which could be adjusted down to the level required when run to make that particular moulding. A smaller machine of 1,000 kN could not be adjusted above its limit and would be inadequate.

Small mouldings are usually made economically in what are called 'multi impression' moulds so that many are produced at a time, not just one. The total projected area is simply arrived at by multiplying the projected area for one moulding by the number of impressions.

Let us imagine that our moulding is a box 10 cm long x 12 cm wide and therefore has a projected area of 120 cm^2. Note that the height of the box does not come into the projected area calculation. Like the beaker, the box is moulded with its bottom towards the injection unit and the top edges of the sides towards the clamping unit and only the length and width are relevant. Whatever the height may be, it does not affect the projected area. Applying the approximate formula and taking 6.2 kN per cm^2 we obtain 744, say 750 kN, as the clamping force required : ($120\ cm^2 \times 6.2\ kN = 744$). The actual margin of error with this method may be +/- 40 per cent but any miscalculation will come to light at the stage of arranging for a mould trial on a machine. The machine manufacturer will apply a better formula to check the calculation, as so should, in due course, the processing technician.

2. Mould width.

This is the horizontal distance from one side of the mould to the other. The dimension will be shown on the drawing for the mould or may simply be measured on the mould itself. Here we will assume a mould width of 225 mm.

3. Mould length.

Length is the tallness of the mould and a vertical dimension. To the unitiated it would appear that it should be the mould height, but it isn't and the reason why is described in the next paragraph. It is necessary to consider this dimension only when the mould is exceptionally long, as will be explained, but we will assume 525 mm, which is moderate, in this case.

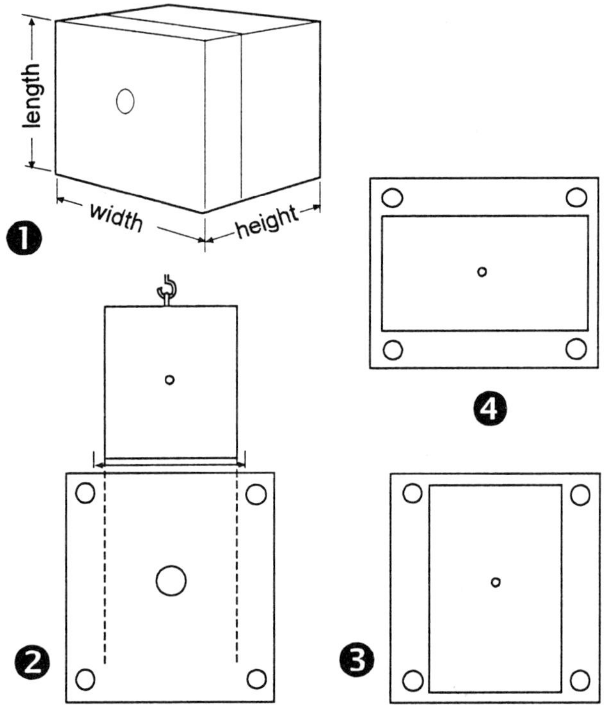

1. The width, length and height of a mould. Note that height is a *horizontal* dimension and length *vertical*..
2. The horizontal distance between the tiebars of the machine should be greater than the actual mould width to allow it to be lowered in between them. Lowering is the usual practice, but if the mould is entered from the side the vertical distance between the tiebars becomes relevant. On most machines the horizontal and vertical distances between tiebars are similar but on some there is a difference.
3. Overall platen height should be greater than the length of the mould so that it does not protrude at top or bottom.
4. Some makers offer wide platen versions of their machines for long moulds.

4. Mould height or thickness.

These terms are synonymous but the important point to bear in mind is that, perversely, ‘height’ is a *horizontal* dimension. Most presses (another word meaning ‘machines’) used in injection moulding are of the horizontal type but in older processes the vertical type of press is traditional. A vertical press is one in which the moving platen goes up and down. ‘Height’ is a term used for a long time in traditional vertical press technology and got carried over in relation to horizontal presses. When a mould is mounted in a vertical press, its height is just that, but when mounted in a horizontal press it is considered to have been turned through 90 degrees and its height becomes a horizontal dimension. ‘Thickness’ is a preferable term but not generally used because the injection moulding

fraternity has become familar with what 'height' means in this context. This anomaly has to be appreciated at once by a newcomer if he is to make any sense of the machine clamping unit specifications.
Some manufacturers refer to this dimension as the mould build height, and an American term is mould shut height.
There is a possibility of further confusion with the term 'mould height' because it is applied both to the mould itself and the machine but has a slightly different meaning in each case. The height of the mould itself and the machine mould height are different concepts and a distinction must be made here to remove a possible obstacle to understanding. In the following text either the term 'mould height' or '*machine* mould height' is used, as appropriate. 'Mould height' simply means the thickness of the mould, as described above. 'Machine mould height' means the distance between the platens allowed for the size of the mould. There is a range of adjustment to suit actual 'mould height'. The reason for making the distinction should become clear.
For the present purpose of explaining how to read a machine specification table, we will assume a mould height of 275 mm.

5. Opening stroke required.

This is the the distance needed to get the moulding out of the mould. It is twice the height of the moulding (also called the 'part' or the 'component') plus an allowance for clearance so that it can either fall out by gravity, be extracted by a robot or taken out by hand. Sufficient space should be allowed for the entry of a takeout robot or, in the case of manual extraction, for the operator to reach into the mould space (the space between the machine platens). Here we will assume that an opening stroke of 325 mm is needed.

▲ Examining machine specifications.

Having written down our requirements 1 to 5, we may now look at a manufacturer's table of machine specifications. This will have a list of about 20 or more machine dimensions and other values but only about half a dozen of these are relevant at this stage of initial short listing. Here we will look only at these few for the moment and deal with the others later. Our list of minimum requirements is like this:

Clamping force	750 kN
Mould width	255mm
Mould length	525mm
Mould height	275mm
Opening stroke	325mm

1. Clamping force.

We first take the line in the manufacturer's table entitled 'Clamping force kN'. This will be under the sub-heading 'Clamping unit' and may, for the purposes of explanation, list the following sizes of machine: -

Clamping force kN	250	350	500	800	1,000

Here we want 750 kN so that the 250, 350 and 500 kN machines can be ignored and the 800 and 1,000 kN models considered. We have now taken our first and most important step. We are looking for an 800 kN or possibly a 1,000 kN machine, that is if either meets all the other requirements. As already mentioned, clamping force on a machine is adjustable. The figure stated in the specification is the maximum and an 800 or 1,000 kN machine can be set to provide only 750 kN.

It is often the case that the physical size of the mould has a greater bearing upon which machine to choose than clamping force and one would therefore first consider points 2 to 5. But here we will follow the classic method. The same end result will be reached and the route taken will be hardly any longer.

2. Mould width.

The table is next examined in respect of the width of the mould, where we are looking for a minimum of 225 mm. The relevant line in the table to look at is called 'Distance between tiebars' and may contain the abbreviations H and V.

H = horizontal distance between tiebars.

V = vertical distance between tiebars.

Where neither 'H' nor 'V' is stated, the first figure given in the specification is the horizontal distance and the second the vertical. Many machines have square platens with equal H and V distances. Some makers also give the diagonal distance between tiebars, which is relevant to mounting cylindrical moulds.

The most common moulding shop practice when fixing a mould in a machine is to lower the mould between the upper tiebars either from an overhead travelling crane or a portable floor crane. A fundamental question is therefore whether the the upper tiebars are wide enough apart for this to be done.

Having eliminated the smaller machines from our selection because they have insufficient clamping force, the 800 and 1,000 kN machines remain and the line

'Distance between tiebars' may read like this:

	800 kN	1,000 kN
Distrance between tiebars mm HxV	305 x 280	400 x 400

The width of the mould is checked against the H distances between tiebars. As can be seen, these are 305 and 400 mm respectively. Allow for adequate clearance so that the mould is not likely to swing against and strike the tiebars as this could cause damage. How much clearance is a matter of common sense but ideally 25 mm on each side is probably the tightest practical limit for a small mould. In practice, this advice is often hardly observed, but to disregard it may lead to damage to the tiebar and mould. The temptation to fit a mould in a machine with a tight clearance arises from the fact that a larger machine would be more expensive. Also, the injection unit on a larger machine might not be of a suitable capacity, that is it could be too big for the mould in question. Moulders therefore try and squeeze a mould in on a machine with the most suitable injection unit even if it has only barely adequate tiebar clearance. As suggested, this calls for considerable care. The allowance needed for large moulds will be greater. Any scratch or dent made accidentally to a tiebar by knocking a heavy mould against it during loading will in turn damage the bearing in the platen as it travels to and fro over it during the cycling of the machine. Blankets or other pieces of material are usually hung over the tiebars for protection and it is sensible to allow space for these.

At least 50 mm (25 mm on each side) should prudently be added to the actual width of a mould intended for a machine in the 800 to 1,000 kN range. Our mould here is 225 mm wide and so we must take 275 mm. The 800 kN machine, with 305 mm, can accept a mould of this width, and so can the 1,000 kN with 400 mm. But the 800 kN machine would have to be ruled out for a mould more than 255 mm wide as clearance for mounting it would either be uncomfortably tight or not exist at all. It is, however, possible to remove an upper tiebar to mount a large mould, but this will involve some downtime, although some large machines have a special feature for this to be done under power.

If the width of the mould is greater than the distance between tiebars, it may still be possible to load it if its height is less than this distance, as may be the case with moulds for flattish products. This is simply achieved by turning the mould through 90 degrees, lowering it between the tiebars and then returning it to the correct orientation.

Some manufacturers offer two sizes of platen, and two sets of tiebar distance figures, with each clamping unit of a certain kN force. In this case, it may be possible to find a suitable tiebar distance for a larger than average mould and avoid seeking a clamping unit with greater kN force than necessary. Normally a machine with sufficient clamping force will provide enough distance between tiebars for most applications and only unusually large moulds present a problem in this respect. Where a clamping unit with a higher than necessary clamping force has to be chosen simply on account of the distance between tiebars that it provides, the situation has to be accepted, and this in fact is often the case.

A choice of platen sizes with similar kN clamping force in a manufacturer's range not only helps to solve problems of selection in respect of unusually large moulds but also moulds which are smaller than average. The smaller of two available platen sizes is a better choice for a small mould as this will be better supported on the platen. The clamping force will be concentrated where most needed.

If the mould is entered into the machine sideways with a fork truck or table type lifting truck rather than from above, the vertical distance between tiebars has to be checked and an addition made to the dimension of the mould for the depth of the forks or the lifting table.

Tiebarless machines are available from certain manufacturers and the matter of distance between tiebars does not arise with these. Some small machines have only two tiebars and there are also three tiebar machines for specialist classes of moulding, notably Compact Discs. There are automatic and semi automatic side loading systems for moulds. These are known as AMC (Automatic Mould Change) and QMC (Quick Mould Change) systems and run the mould in from a roller table placed at the side of the machine at a level slightly above that of the lower tiebars. Selection of a machine with such a feature would require special consideration vis a vis vertical tiebar distance as large mould backplates are generally fitted. AMC and QMC systems are also available for loading moulds from above.

3. Mould length.

We now come to the third point, the length of the mould. The overall size of the platen is given in the specification table and may be relevant where moulds of exceptional length or width are involved. Ideally the mould should fit within the platen for good clamping support and not protrude at top and bottom or at the sides. Our imaginary mould is 525 mm long and the overall platen size for the 800 kN machine is given in the table as 590 x 580 mm HxV. Mounted upright on the platen, the mould will fit with 55 mm to spare, which may be considered sufficient but not particularly generous. If inadequate for good mould support, the figures for overall platen size for the next model up in the range would have to be examined. Some manufacturers offer optional oversize platens for long moulds such as those for car bumpers or fluorescent light diffusers.

4. Mould height / 5. Opening stroke required.

Consideration of the next points, mould height and opening stroke, is more complicated. Because of the basic difference between toggle and hydraulic machines, minimum and maximum machine mould heights are stated in a specification for a toggle machine but only the *minimum* machine mould height for a direct hydraulic machine, *maximum* machine mould height being omitted.

Why this is so requires a little technical explanation (see panel below). A toggle machine has a relatively small hydraulic cylinder and the cylinder ram acts against the moving platen via a fairly complex 'lazy tongs' toggle mechanism. A direct hydraulic machine has a much larger hydraulic cylinder and the ram acts directly against the moving platen.

Points 4 and 5 need to be considered separately in relation to toggle and direct hydraulic machines.

Toggle machines.

The major part of the clamping unit assembly of a machine of this type, comprising the tailstock, toggle system and moving platen, is steplessly adjustable over a certain distance on the base frame and can be fixed in a chosen position. Steplessly means that the adjustment can be anywhere as distinct from any one of a few predetermined incremental points over the range of adjustment. ('Steplessly' is now used instead of the term 'infinitely'.) The actual mould height dimension should be greater than the minimum machine mould height, and less than the maximum machine mould height. In the first instance the machine would otherwise be too big for the mould and in the second the mould would be too big for the machine. Some manufacturers offer non standard minimum and maximum machine mould height variations to suit special applications so that unusually short or high moulds can be accommodated. In our example, the actual mould height is 275 mm and the minimum/maximum machine mould height adjustment range shown in the table is 200 to 450 mm. There is therefore no problem here as 275 mm comfortably exceeds the minimum on the machine.

An opening stroke of 325 mm is required in order to extract the moulding and the amount of stroke available on the 800 kN machine is shown in the table as 420 mm. Again this is adequate and it is unnecessary to look at the specifications for a larger model.

Direct hydraulic machines.

As mentioned above, only a minimum machine mould height is stated in the table for such types. With the exception of a few models of old design, direct hydraulic machines have no separate adjustment for mould height as the tailstock with the clamping cylinder is fixed to the frame and is not adjustable. The opening stroke of a direct hydraulic machine, however, is normally longer than for a toggle machine of equivalent kN clamping force and this provides a margin for machine mould height adjustment.

The minimum machine mould height shown for an 800 kN direct hydraulic machine is 250 mm and this is acceptable as the actual mould height, 275 mm, exceeds this. The amount by which the actual mould height (275 mm) exceeds minimum machine mould height (250 mm) is 25 mm and this must be deducted from the figure given for the machine opening stroke. The stroke shown in the table is 450 mm and this therefore comes down to 425 mm. This distance is compared against the amount of opening stroke actually required. In our example the opening stroke required is 325 mm, so that the machine is adequate in this respect.

Minimum and maximum mould height and opening stroke

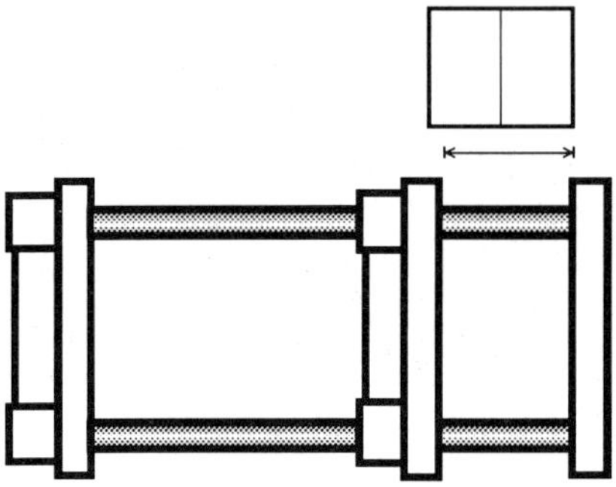

❶ Minimum mould height of the machine must be less than the actual mould height.

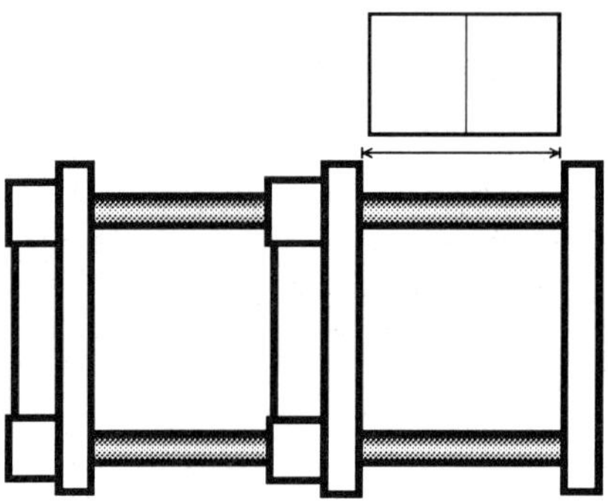

❷ With a toggle machine and a mould of unusually great height, the maximum machine mould height must be greater than the actual mould height.

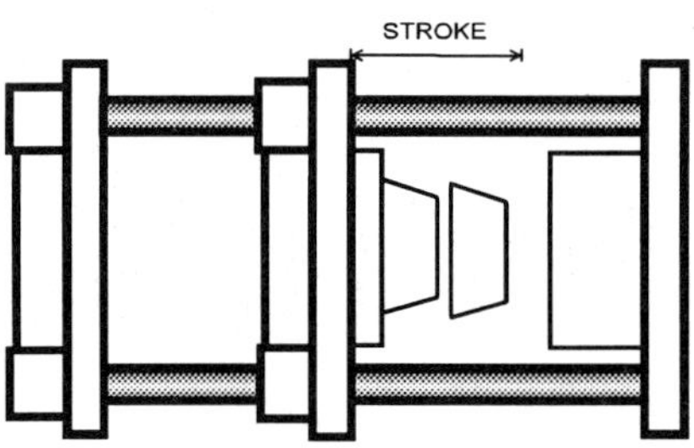

❸ The opening stroke must be long enough to allow the moulding to be extracted from the mould. On a direct hydraulic machine, the amount by which actual mould height exceeds minimum machine mould height reduces available length of stroke and this must be taken into account.

The term 'direct hydraulic' does not imply that a toggle machine is not also hydraulically actuated, solely that the actuation is direct. Direct hydraulic machines are sometimes referred to simply as 'hydraulic' machines, or 'direct ram' machines. Toggle and direct hydraulic machines between them account for the majority of models sold, but there are also hydro mechanical, or 'lock and block', types. The specifications for most hydro mechanical machines read like those for toggle machines as they also usually have a range of machine mould height adjustment. Hydro mechanical machines are chiefly large models from 4,000 kN upwards.

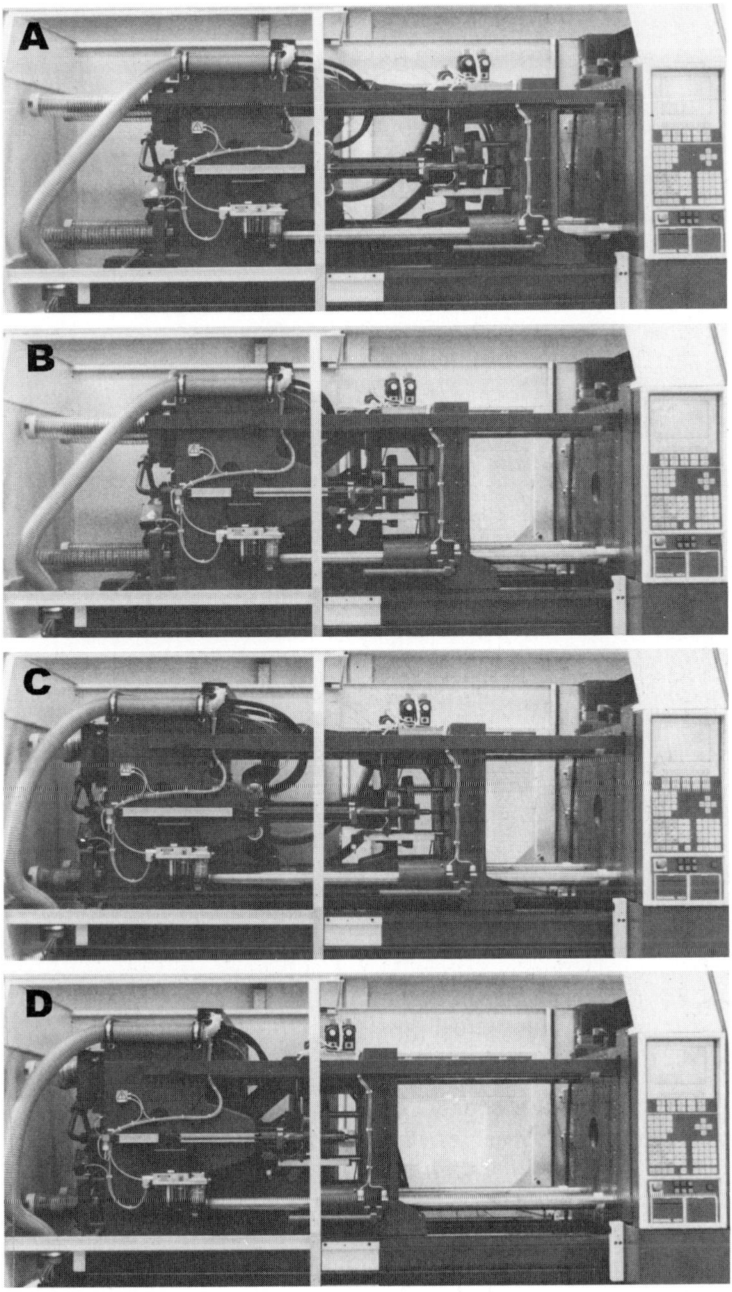

These pictures illustrate mould height adjustment and mould opening stroke. Pictures A, B, C and D show a toggle machine and E and F on page 38 a direct hydraulic type. See captions overleaf.

A.

In A, the tailstock has been moved forward through a range of adjustment on the frame of the machine and the toggle system is extended. The distance between the fixed and moving platens is said to be the 'minimum mould height'.

B.

Picture B shows the toggles fully retracted and the moving platen brought back. The distance through which the moving platen has travelled is called the 'stroke'. This movement is made to open the mould so that the moulding can be taken out.

The minimum mould height as shown in A would not be enough for larger moulds but the machine can be adjusted for these.

C.

It can be seen in C that the tailstock has been brought back almost to the end of the frame. The toggle system is fully extended and the distance between the fixed and moving platens is greater than in A. This provides the 'maximum mould height'. Any intermediate adjustment between minimum and maximum mould height is possible.

Stroke length is unaffected by mould height adjustment.

D.

In D, the toggle system is retracted and the moving platen has travelled back for the same distance as in B. The distance between the platens in D is termed the 'maximum daylight'. The length of stroke actually used is according to how much is needed to extract the part from the mould. The full stroke length available need not be used for small or medium sized parts.

With most direct hydraulic machines, the tailstock is fixed to the frame and therefore has no range of adjustment as on a toggle machine. Instead the direct hydraulic machine has a comparatively longer stroke to provide for mounting moulds of different sizes whilst still allowing adequate length of stroke for moulding extraction.

E.

Picture E shows a machine of this type with the ram fully extended and the moving platen forward in the minimum mould height position.

F.

In picture F the ram is retracted and the moving platen has travelled back through its full stroke and is in the maximum daylight position. The amount by which the actual height of a mould exceeds the minimum mould height reduces the length of stroke that can be used, but a direct hydraulic machine is designed to allow for this. The difference between the two types of machine arises because on a toggle machine it is necessary for the toggles to be fully extended for clamping force to be applied, whereas clamping force on a direct hydraulic machine can be applied whether or not the moving platen is at the extreme end of its forward stroke.

Special machines are built by some manufacturers where there is a high volume of demand for the mouldings concerned. For example, a dedicated machine to mould Compact Discs has 60 tonnes of clamping force but an opening stroke of only 60 mm, which is adequate for such a thin moulding but little else. Even so, 60 mm opening stroke at first sight seems generous for a moulding only 1 mm thick. The reason is that the mould opening stroke has to allow for the entry of a take out robot.

It would help buyers if more manufacturers gave in their tables of specifications an indication of the projected area of moulding possible according to clamping force.

For example, one manufacturer states in his specifications for a machine with 1,865 kN of clamping force that the maximum projected area is 746 cm^2, on the assumption of a cavity pressure of 250 bar. This figure is far more helpful than some others traditionally included in a typical table. The diagram in the Appendix (page 171) shows how the amount of cavity pressure relates to wall thickness and length of flow path.

Let's recapitulate here by recalling our list of requirements and comparing them against the figures for the hypothetical 800 kN machine we have selected in toggle and direct hydraulic versions:-

Requirements	Machine specifications
1. Clamping force 750 kN	800 kN
2. Mould width 225 mm.	Distance between tiebars 305 x 280 mm HxV
3. Mould length 525 mm.	Overall platen size 590 x 580 mm HxV
4. Mould height 275 mm.	Min/max mould height (toggle machine) 200/450 mm
	Min mould height (direct hydraulic machine) 250 mm
5. Opening stroke 325 mm.	Opening stroke (toggle machine) 420 mm
	Opening stroke (direct hydraulic machine) 450 mm

We may just go through these figures again to round off and reiterate the description of the process of selecting a clamping unit.

1. The clamping force required, 750 kN, is within the capacity of the machine as this is 800 kN.

2. Assuming that the mould is loaded into the machine from the top, a horizontal distance between tiebars of 225 mm plus an allowance for clearance is needed. The machine has 305 mm horizontal distance between tiebars, so is adequate.

3. Mould length has to be considered in relation to overall platen size. The actual mould length is 525 mm and the overall vertical dimension of the platen 580 mm, giving a margin to permit the mould to be mounted without projecting at the top or the bottom.

4. Actual mould height is 275 mm. This is greater than the minimum mould height of both the toggle and direct hydraulic machines and these are therefore satisfactory.

5. The opening stroke of 420 mm for the toggle machine exceeds the opening stroke actually required, 325 mm, so there is no problem. In regard to the direct hydraulic machine, the actual mould height exceeds the machine mould height by 25 mm. This distance must be deducted from the machine's opening stroke. This leaves an effective 425 mm, which is adequate.

▲ Choosing the injection unit

We now need to find the most suitable injection unit. Write down the figures for the following in respect of the moulding. A figure for Point 6 should be known from the original specification for the mould and the part. Point 7 will be explained: -

6. Shot volume cm^3 / 7. Injection pressure bar.

6. Shot volume cm^3, or Shot weight g if shot volume is not known.
7. Injection pressure bar. (Bar is the unit of pressure, as in 'barometer'.)

6. Shot volume.

Look lower down the table from the section dealing with the clamping unit to that for the injection unit. A line of figures will be given in cm^3 for shot volumes and manufacturers also give figures in (g) for shot weight. Shot volume is the more precise expression of the two. The figures for shot volume in the table are the maximum volumes of material that can be injected. Some manufacturers give what they call the theoretical shot volume, which is the maximum screw stroke distance multiplied by the screw cross sectional area, and others a figure less an allowance, usually 10 per cent, for backflow of material over the front end of the screw on injection. For example, if a 100mm diameter screw has a maximum stroke of 500 mm, the theoretical shot volume will be given as 3,927 cm^3, that is the total volume swept by the stroke. Another figure, which may be called the actual shot volume and takes a 10 per cent loss through backflow into account, might also be shown and in this instance would be 3,534 cm^3. This can be taken as the truer figure for practical purposes.

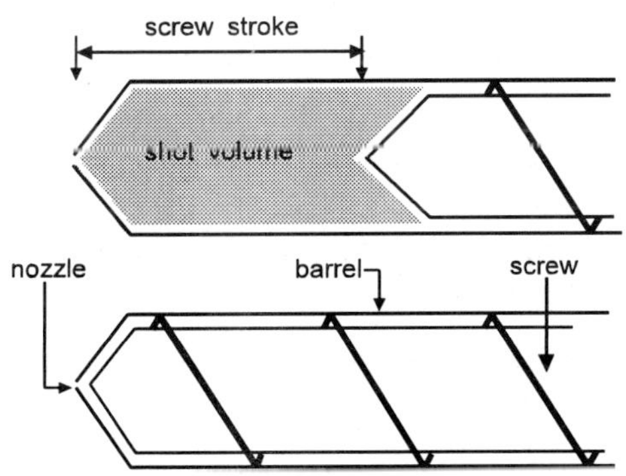

Shot volume is depicted in the simplified diagrams above. As well as being able to rotate, the screw travels backwards and forwards over a certain stroke distance and is therefore said to reciprocate. The backwards stroke is performed when it is rotating and plasticising the material. Material accumulates at the front, as shown by the shaded area in the upper diagram, during this phase. The screw then travels forward (without rotating) to force the material out through the nozzle and into the mould.

The maximum shot volume figure is shown in the machine specification but the amount to use in actual practice must be:-
(a) at least 20 per cent of the maximum shot volume and preferably 40 per cent, and
(b) no more than 80 per cent and preferably 60 per cent of the maximum shot volume.
Whether it is (a) or (b) depends upon the quality of moulding required.
Selecting an injection unit is therefore not a matter of choosing one with a maximum shot volume to match the volume actually required for the moulding almost exactly. The amount required for the moulding should be between 40 and 60 per cent of the volume as shown in the shaded area for high quality production or between 20 and 80 per cent for less demanding work.
Machines are designed to give certain shot *volumes* rather than weights. Shot weights derive from volumes but are sometimes given for the guidance of prospective customers who know the weight but not the volume of the moulding concerned. Preferably one should know the shot volume rather than the shot weight, but this is not always the case. Shot volume will be discussed first.
As mentioned above, for a model of a given clamping force in kN, the manufacturer usually offers two or three injection units with different ratings. (Euromap injection unit ratings are described on page 51.) There is usually a choice of two or three screws of different diameters for each injection unit and the one that combines sufficient shot volume with adequate injection pressure needs to be found.
As an example, let us look at figures for a typical 800 kN machine as they might appear in a manufacturer's specification:-

Clamping Unit kN	800								
Injection unit rating	200			310			430		
Screw diameter mm	25	30	35	30	35	40	35	40	45
Shot volume cm^3	61	106	144	106	168	220	168	231	293
Injection pressure bar	2400	1990	1460	2400	2020	1550	2400	2020	1600

This 800 kN machine is available with three injection units rated at 200, 310 and 430 respectively and there is a choice of three sizes of screw (in terms of diameter) for each injection unit. We, however, are more interested in the shot volumes rather than the screw diameters. Let us imagine that we are looking for a machine with a usable shot volume capacity of 100 cm^3. Do we choose the 30 mm diameter screws available with the 200 and 310 rated injection units? It might appear so as each gives a shot volume of 106 cm^3, apparently enough as 106 exceeds 100 by a slight margin, but matters are a little more complicated than might be presumed and the answer is 'No'.
The shot volume of the machine should be greater than that represented by the

moulding to ensure satisfactory thermal characteristics in processing (see page 93). Most machinery manufacturers and other experts on injection moulding recommend that no more than 80 per cent of the available shot volume of the machine should in practice be used, or better still 60 per cent. It is not advisable to use all the available shot volume of the machine.

If the volume of the moulding plus an allowance if necessary for a sprue and runner system is 100 cm^3, for example, this figure must represent no more than 80 per cent of the shot volume capacity of the machine. This means increasing the figure for the volume of the moulding by 25 per cent because 100 cm^3 is 80 per cent of 125 cm^3. One then looks along the line 'Shot volume' until a figure of at least 125 cm^3 is given.

However, 125 cm^3 may be regarded as a borderline practical minimum and it would be more satisfactory if 100 cm^3 was only 60 per cent of the shot volume available. This is the maximum percentage recommended for moulding to a high standard of quality. In this case, we increase our 100 cm^3 by 66 2/3per cent, say 67 per cent, and therefore look in the table for a shot volume capacity of 167 cm^3. We are now therefore looking for a shot volume capacity of at least 125 cm^3 but preferably 167 cm^3.

It would probably be better if manufacturers did not give maximum shot volumes but figures for 60 per cent or 80 per cent of the maximum as these are more realistic than 100 per cent. So long as they do not a prospective buyer must interpret the figures given in a specification table according to his own requirements and must certainly not be misled into thinking that the shot volume required and the actual shot capacity of the machine should be similar. A good rule of thumb is to assume that the shot volume of the machine should be twice that of the moulding. Thus if the moulding has a volume of 100 cm^3, a machine with a shot capacity of 200 cm^3 may be selected.

There is also a *minimum* proportion of the shot volume which may advisably be used (again see page 93: Proportion of total shot volume which should be used) but obviously it is the maximum which needs to be reckoned with in selecting the most suitable size of screw at this stage.

If we then look at the table, it will be seen that the 35 mm screw of the 200 injection unit has a shot volume of 144 cm^3 and so exceeds the minimum requirement of 125 cm^3 by a small margin. But the same size of screw on the 310 and 430 injection units just meets the more desirable requirement for 167 cm^3 by providing 168 cm^3. If the 35 mm screw of the 200 injection unit can be accepted, we have therefore arrived at the stage where the required injection pressure can be considered.

It may be wondered, from looking at the tables, why the 35 mm screw of the 200 injection unit gives only 144 cm^3 shot volume whereas the same size of screw on the other injection units gives 168 cm^3. This is because the screw stroke on the 200 unit is shorter than on the others and less material can be injected. Screw stroke distances are sometimes given in a specification table and in this example the stroke of the 35 mm screw on the 200 injection unit is 150 mm compared with 175 mm on the 310 and 430 units. There may also be differences in stroke lengths between screws on the same injection unit, but the prospective buyer need not be primarily concerned with stroke lengths as it is the shot volume and injection pressure figures which chiefly matter to him. The injection pressure figures quoted in a table are the maximum available on the machine. Actual pressures which may be used are adjustable up to the maximum.

7. Injection pressure.

Broadly, injection pressures for common 'commodity' materials are between 500 and 1,000 bar and for technical engineering materials 800 and 1,600 bar. The material manufacturer also provides other relevant data including the maximum flow path length it is possible to mould for a certain wall thickness. The very first step to take when considering making a part in plastics, even before designing it and proceeding to make a mould and select a machine, is to select the type and grade of material in which to make it, and the relevant data should be carefully studied.

Because the molten plastics may have to be forced through a number of constricted passages on its way to the cavity, and in the cavity possibly have to flow around corners and through narrow gaps, the pressure at which it starts its journey in the barrel, that is the injection pressure, is considerably reduced by the time it flows into the mould cavity and reaches the end of its flow path. For example, if material is injected with an injection pressure of about 1,200 bar, a pressure of only 250 bar may be attained in the cavity because of the losses in pressure en route.

In practice, the injection pressure for a particular mould is found by deliberately starting with a low shot volume and inspecting the moulding produced. This will be only partially complete because the cavity has not been wholly filled. The volume is then carefully and gradually increased stage by stage and the mouldings examined until a satisfactory result is obtained. This can be memorised for future recall on machines with microprocessor control. Caution is called for as there are a few kinds of moulding to which this technique cannot be applied, for example bottle crates. The machinery supplier should be consulted in such circumstances. Some makes of machine have special control programs to find the best injection pressure setting automatically.

Recommended injection pressures will be found in the specifications for the material concerned published by the materials manufacturer. For example, 800 to 1,200 bar may be recommended for PP (polypropylene) and 1,000 to 1,600 for acrylic. An estimate of where to set injection pressure between the minimum and maximum limits is made according to the wall thickness of the moulding and the flow path length (see appendix page 171) and this is confirmed later during a mould trial on a machine.

If 1,600 bar, for example, is the required injection pressure for the moulding, the 35 mm screw of the 200 injection unit is inadequate as it gives a maximum of only 1,460 bar, but enough pressure is available on the 310 injection unit at 2,020 bar, and even more on the 430 at 2,400 bar. The 35 mm screw of the 310 injection unit may therefore be chosen because going up to the 430 unit is unnecessary.

The 310 unit with 35 mm screw gives the optimum combination of shot volume and injection pressure.

For the great majority of mouldings, however, the maximum injection pressure required in practice is below 1,000 bar. This usually means that the second smallest diameter screw offered with an injection unit of any particular rating will be suitable. The high maximum pressure of the smallest diameter screw is normally used only for thinwall work, but there is no reason not to choose the smallest diameter screw if its shot volume is acceptable. It can simply be run at a pressure setting below its maximum if this is not required.

Where traditional moulding methods are employed, that is without a hot runner system, a certain amount of plastics material is necessarily used simply to fill the channels in the mould leading to the cavities, or 'impressions', in which the mouldings are actually formed. A 'sprue' is formed in the entry channel in the mould in line with the nozzle of the machine and 'runners' in any channels branching off it to individual cavities. Although they do not form a part of the moulded products, they have to be taken into account when reckoning the shot volume required. This must therefore cover the sprue and runners plus the parts themselves. It is not uncommon, when small parts are moulded in a multi cavity tool, for the sprues and runners to account for more material than the parts themselves.

Sprues and runners are cut off either automatically or by hand when the moulding is ejected. They are then chopped up in a machine called a granulator for reuse. What are known as 'hot runner systems' enable mouldings to be made without sprues and runners and only the shot volume represented by the mouldings need be considered. Hot runner systems maintain temperature in the sprue and runner channels by means of electric heaters to keep the material in a molten state. Material enters the mould cavities through extremely small 'pin' gates and there may be an on/off valve at these points to control the melt flow.

Shot weight.

Shot weight, as distinct from shot volume, may now be considered, but remembering that it is preferable to read a table with shot volume in mind. Often, however, the weight of a moulding will be known but not the volume, presenting a slight problem in choosing a suitable injection unit. Some manufacturers help a little towards solving this difficulty by quoting shot weights in polystyrene (PS), one of the more widely used injection moulding materials. In such a case it will be noted that the figure shown in grams in the specifications is about 10 per cent less than the figure for shot volume in cm^3. This is because the Specific Gravity (that is the weight in grams per cm^3) of polystyrene is 0.91. However, with the exception of nylon (PA), the SGs of other materials are not the same as for PS and, where given, the shot weight in this material would not be a true indication of that in another.

A method of calculating shot volume required from the known weight of a moulding is described on the table on the next page. One column of this table

gives the SG figures for some popular injection moulding materials in their molten state, and another the reciprocals of these figures. The reciprocal of a number is 1 divided by that number, for example the reciprocal of 2 is 0.5.
To use the table, first refer to the type of material, then read off the reciprocal. Then multiply the weight by the reciprocal. (Do not forget the weight of the sprue and runners, if applicable.) This will give the shot volume.

	SG	Reciprocal
ABS (acrylonitrile butadiene styrene)	0.88	1.14
Acrylic - PMMA (polymethyl methacrylate)	0.94	1.06
LDPE (low density polyethylene)	0.76	1.32
HDPE (high density polyethylene)	0.73	1.37
Nylon - PA (polyamide)	0.91	1.10
PC (polycarbonate)	0.97	1.03
PP (polypropylene)	0.73	1.37
POM (acetal)	1.15	0.87
PPO (polyphenylene oxide)	0.84	1.19
PS (polystyrene)	0.91	1.10
uPVC (unplasticised or rigid polyvinyl chloride - also referred to as PVCu)	1.02	0.98
SAN (styrene acrylonitrile)	0.88	1.14

The SG figures in this table and their reciprocals are for the materials not in their normal solid state as granules at ambient temperature but as a melt. There is a considerable difference between the SG of a material in its solid and molten states and the figure for the molten state must be taken. This is because the material exists not as a solid but a melt in the barrel ahead of the screw tip, that is in the part of the machine to which the shot volume figure relates. For example, PC (polycarbonate) has an SG of 1.20 in its solid state but 0.97 as a melt and a considerable error could be made in reckoning shot volume if the figure for the solid state were taken. As a melt, the material has been expanded with heat and is considerably less dense than in its solid condition and its SG is lower.
Looking at the table above, it can also be seen that PC has a higher SG than SAN, 0.97 as against 0.88. Therefore a smaller machine shot volume is required to produce a moulding of a certain weight in PC than in SAN. This is reflected in

the respective reciprocal figures - 1.03 for PC but 1.14 for SAN. If a PC moulding weighs 150 g, the amount of shot volume that represents is 154.4 cm^3 (150 x 1.03), but a moulding in SAN weighing the same would need a shot volume of 171 cm^3 (150 x 1.14).

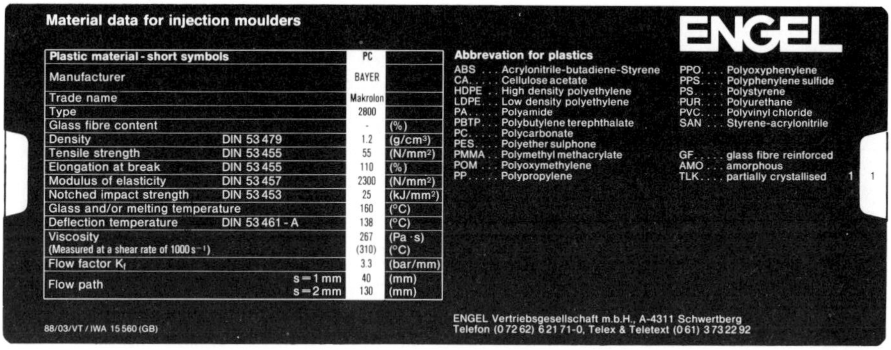

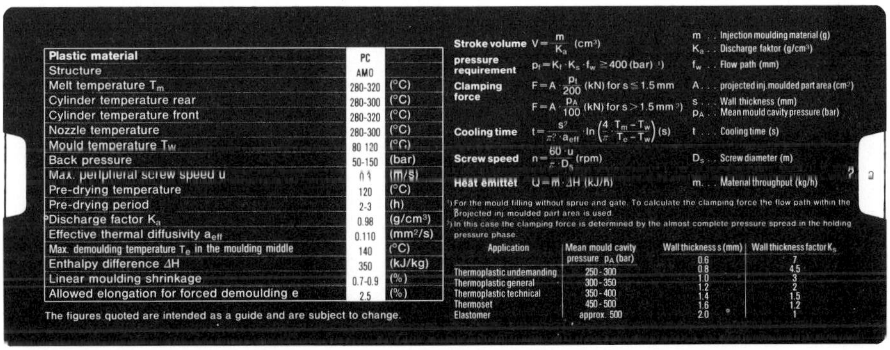

This special slide rule enables cavity pressure, clamping force and shot volume to be calculated for mouldings in 21 different materials. Cooling time and screw speed can also be reckoned. Factors taken into account include the Specific Gravity of the material in its molten state (called the 'Discharge Factor') and the Flow Factor, the amount of pressure needed for each 1 mm of flow path length. These vary according to type of material and lists of values are given for these. The slide rule is available from Engel UK. (Address given in 'Further reading' section.)

It is the reciprocal figures in this table rather than those for SG which are useful. The reciprocal of a number can easily be found by using a pocket calculator. As stated, to find out the shot volume for a moulding in a particular material, the known weight of the moulding is multiplied by the reciprocal for the type of material.

Let us see how this works out. If a moulding in polypropylene weighs 80 g, 80 is multiplied by 1.37, as given in the table. This gives 109.6, which is the machine shot volume required in terms of cm^3 - but, it must be emphasised, only the starting point, for it remains necessary to adjust this figure by going through the steps described above. The shot volume required for the moulding should not exceed 80 per cent, or better still 60 per cent, of the shot volume available on the machine.

Reciprocals for other materials can be added to this list as desired. Materials manufacturers should be able to give the SGs for their materials in the melt state as the basis for simple calculations. Note the words 'in the melt state'. SG figures relating to the normal solid state would produce meaningless results. However, it would be unrealistic to pretend that great accuracy is required for preliminary calculations of this kind. Trials with a machine and mould must always be made to verify theoretical figures. The object of calculating shot volume figures from known moulding weights is to lead oneself broadly but with fair accuracy in the direction of the most suitable injection unit. Attaining ultimate exactitude depends upon a number of other factors.

chapter 2

Other headings in a specification table explained.

The most critical factors in a machine specification have been considered in Chapter 1 and any reader wanting only an essential briefing on the subject can finish here, if he wishes. The full list of features as given in a manufacturer's specification will next be discussed, followed by a review of a typical list of optional extras in the next chapter.

The modular design of modern injection moulding machines, which involves an initial overall conception by the manufacturer of a range of models in a number of clamping unit sizes and shot volume capacities, permits a buyer to pick and choose individual parts and put together on paper the machine which best matches his requirements. It may be seen from a table of specifications, for example, that an injection unit of a certain rating and with a choice of 30 and 35 mm diameter screw sizes is available on a 500 kN machine as well as an 800 kN. This makes it possible to combine the more suitable size of screw with either of the two sizes of clamping unit. Manufacturers are also normally willing to depart from standard specifications where possible and combine modular assemblies in ways which they had not anticipated, for example providing a machine with a certain clamping force but platens of a size specified for the next heavier model in the range.

A machine defined in such a methodical way will be the most economical both in terms of initial outlay and subsequent running costs, so, clearly, a great deal concerning the profitability of any injection moulding operation hinges upon careful and systematic selection in the early stages. There is a great distinction between a considered approach to machine selection and merely choosing between different machines on the basis of price or soonest delivery. To be forced into choosing in this manner in effect deprives the buyer of the option of exercising careful selectivity and he is manoeuvred into accepting only whatever might be on offer at a random moment. This might not be the machine best specified for his purpose.

All manufacturers of injection moulding machines set out their technical specifications in much the same form, although the terminology used may vary

slightly from one maker to another, as may the order of the various headings. Also, some makers give certain less important details that others do not.
The first column on the left of a table may contain a list of headings like this: -

Maker's model designation
Euromap size classification
Clamping unit rating
Clamping force kN
Locking force
Overall platen size HxV mm
Distance between tiebars HxV mm
Tiebar diameter mm
Minimum mould height mm
Maximum mould height mm
Maximum mould opening stroke mm
Maximum daylight mm
Hydraulic ejector stroke mm
Ejection force kN
Retraction force kN
Injection unit rating
Screw diameter mm
Length/diameter (L/D) ratio
Injection stroke mm
Maximum injection pressure bar
Shot volume cm^3
Shot weight (PS) g
Rate of injection g/sec
Maximum injection speed mm/sec
Plasticising rate g/sec
Screw rotation speed rpm
Maximum screw torque Nm
Nozzle contact force kN
Carriage stroke mm
Hopper capacity kg
Oil tank capacity litres
Pump kW
Cylinder heating power kW
No. of heating zones
Total installed power kW
Line pressure bar
Dry cycle rate per hr
Net weight kg
Dimensions mm

The meanings and significance of these headings are now discussed individually, but the reader must be given a preliminary warning that this necessarily involves going into some of the technicalities of injection moulding. The text is broken up with separate headings so that he may skip parts of it if he wishes.

A table for a range of machines is normally arranged in vertical blocks across the page or pages starting with the smallest model on the left, for example a 250 kN type, then a 500 kN, an 800 kN and so on. At the top of each block is the manufacturer's designation for the particular model and under this is its Euromap international size classification. These two terms may be dealt with together.

▲ Maker's model designation and Euromap size classification.

At one time, manufacturers tended to designate their models in either of two ways: some included in the designation a number which represented the clamping force in Imperial tons or Metric tonnes, whilst others coined a figure which was simply a notional or series number that gave no indication of the size or capacity of the machine. The result was that there was no valid way of comparing machines of different makes on the basis of their manufacturers' designations.

It was then thought a good idea to invent a formula for machine size classification which all manufacturers would adopt. The Euromap international size classification thus came about, but a change which it introduced was the use of the term 'kilonewton' (based on the Newton, the SI unit for force) instead of tonnes. As nearly as makes no significant difference, 1 tonne = 10 kN so that, for example, a machine with 80 tonnes of clamping force became described as 800 kN. Euromap is the abbreviation of the title of the European Committee of Machinery Manufacturers for the Plastics and Rubber Industries, based in Switzerland.

Some manufacturers soon adopted the international size classification as the basis for their model designations, and in their machine specification tables the numbers given for 'Model designation' and 'International size classification' are identical. But some other manufacturers have not gone this far, the reason being that the concept of tonnes is still firmly rooted in the minds of many moulders, particularly of the older generation, and the relative novelty of kilonewtons resisted or imperfectly grasped.

All manufacturers want to make it easy to sell their machines and for their model designations to be easily understood, and this is why some have continued to use tonnes rather than adopt kN. There may therefore be two otherwise similar machines of different makes, one of which is has the model designation 500/200, for example, and the other 50/200, yet they both have the same clamping force. The 500/200, although designated according to the international size classification, may even be thought by some to be a 500 tonner. Only by reference to the lines for 'International size classification' or 'Clamping force kN' can one be certain what it is.

Some manufacturers will use a tonnage figure in a model designation which does not exactly tally with the actual clamping force in kN but rounds this off to a higher value. A machine with an actual clamping force of 1,834 kN, for example, will be given a model designation '190' and be referred to as a 190 tonner.

Room for confusion does not end here. In the international size classification the first number denotes the clamping force and the second the injection unit rating (see below). But this is not universally adopted and a few manufacturers give the numbers in the reverse order, with the injection unit rating first and clamping unit kN second. Thus one maker will call a particular machine the 800/310 but another, albeit in the minority, will describe an otherwise similar machine as the 310/800. One must again refer to the line 'Clamping force kN' to check exactly what is meant.

The reason given by those manufacturers who put the injection unit rating first and the clamping force kN second is that this reflects the flow of material in the process, in other words that the material first enters the injection unit and then the clamping unit. This may be said to be logical but is inconsistent with the fact that the clamping force is the starting point of the design of the machine. The designer first determines the clamping force and then designs suitable injection units for it, not the other way round. It is also logical, or at least sensible, that all manufacturers should adhere to a common way of using a formula originally introduced to eliminate confusion, but a certain amount of company pride and individuality is involved here and this ideal state may never be attained.

Although it is preferred by most moulders and others in the industry in the UK at the time of writing, the term 'tonnes' will almost certainly give way to kN eventually, as it has done already on the Continent, and the newcomer to injection moulding may advisably start by assimilating the newer expression.

A letter 'H' included in a Euromap designation indicates that the clamping unit is of the horizontal type, as distinct from the vertical, which would have 'V'. Injection units may similarly be described according to their orientation. The majority of injection moulding machines have horizontal clamping and injection units, and it can be assumed that the unit is horizontal where no letter is given.

We may next consider the second figure in the Euromap designation. The injection unit rating of a machine relates to the total volume of the shot that can be injected and the size of the cylinder, or cylinders, that produce the pressure to force the screw forward to inject material into the mould. Comparing two injection units one of which may be designated 445 and the other 200, the 445 will have a bigger shot volume and a larger diameter injection cylinder. There is usually a choice of injection unit ratings for a machine of any kN clamping capacity and the buyer can select whichever is most suitable for his work.

The diameter of the cylinder determines the maximum pressure available to inject the material. The higher the Euromap rating, the larger the diameter of the cylinder and the greater the pressure available. There is usually a choice of two or more injection ratings for any model of a certain kN clamping force. In addition there is a choice of several screw sizes for each rating. This allows a wide selection of shot volume and maximum injection pressure combinations to meet various requirements.

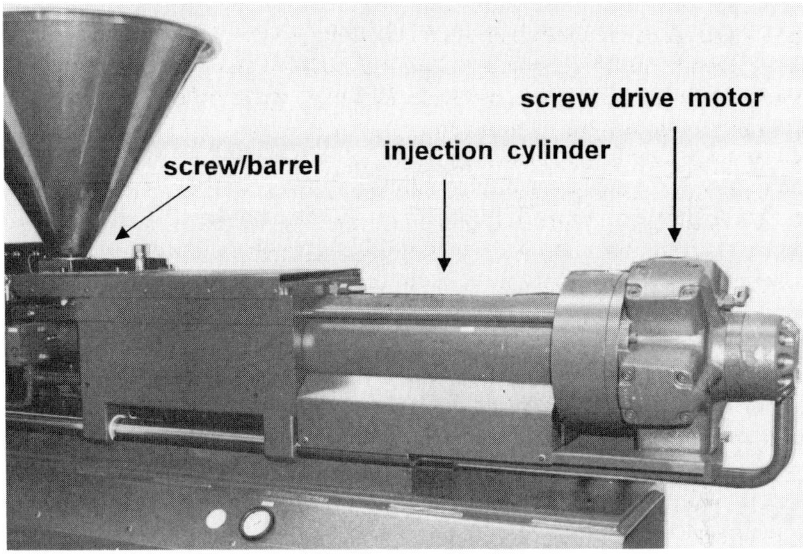

A typical injection unit has a screw drive motor and an injection cylinder arranged as seen here. The motor is driven by the machine's hydraulic pump and rotates the screw to plasticise the material. The piston inside the injection cylinder pushes the screw forward at high pressure to force the melt at the front of the barrel through the nozzle and into the mould.
This machine has one injection cylinder but there are other types with a pair of cylinders arranged one on each side of the barrel and acting on the screw via a cross connecting plate or yoke.

The rating is calculated according to a Euromap formula by multiplying the maximum injection pressure, as shown in the specification, by the volume of the shot and dividing by 1,000. In the case of the particular injection unit rated at 445, the maximum injection pressure for a 40 mm screw is given as 2,219 bar and the shot volume 200 cm^3. Multiplying these two figures together and dividing the product by 1,000 gives 445, or approximately so.
The product of the injection pressure and shot volume is constant irrespective of the diameter of the screw, so that for a larger screw diameter (and hence larger shot volume) the maximum injection pressure available is lower. How this works out is shown by taking the figures for the 45 mm screw also offered with the 445 rated injection unit. The maximum injection pressure is given as 1,748 bar and the shot volume 254 cm^3. Again this gives a rating of 445.
Figures for the alternative 220 rated unit can be taken to illustrate the point further. The cylinder for this is not only smaller in diameter than for the 445 but has a shorter stroke. As the two screw sizes are also smaller in diameter, however, injection at comparable pressures is possible. A 32 mm screw on the 220 unit can inject a 103 cm^3 shot at 2,151 bar, and a 37.5 mm screw 141 cm^3 at 1,565 bar. If the reader works these out for himself he will find that they both check out a rating of 220.
To sum up, the injection unit rating is obtained by multiplying the maximum injection pressure by the shot volume and dividing by 1,000, and this figure is constant irrespective of screw diameter. If screw diameter (and shot volume

correspondingly) increases, maximum injection pressure will be proportionately reduced and vice versa.

This point is relevant as one may buy a machine with more than one screw and barrel, following the example quoted, say a 32 and a 37.5 mm diameter screw for the same 220 rated injection unit. These are exchanged on the machine from time to time as required for different jobs.

Strictly, the Euromap formula for injection unit rating is based on theoretical shot volume at an injection pressure of 1,000 bar. Its chief significance is that as all manufacturers adhere to it there is a comparable basis upon which machine specifications can be assessed. One can take a number of machines of 1,000 kN clamping force, for instance, compare their injection unit ratings and make an initial short listing, having first calculated the rating required. A certain manufacturer's 1,200 kN machine offers two injection units rated at 220 and 445 whilst another maker offers three at ratings of 320, 440 and 610 for a machine of similar kN clamping capacity.

Nevertheless, the Euromap injection unit rating is a compound expression and it remains necessary to check shot volume and injection pressure individually in relation to the requirements for a particular mould. As a guide to machine selection, it is a coarse and not a fine tune and its constituent factors have to be broken down. The value of the Euromap rating does not extend all the way to the final stages of selecting an injection unit. One could have a choice of several injection units with different shot volume/injection pressure combinations but they could all be of similar Euromap rating.

▲ Clamping unit rating.

Some manufacturers have a separate line in their tables for 'Clamping unit rating' figures to provide convenient subgroup headings. The figures in this line reflect the clamping force in kN.

▲ Clamping force kN.

This has already been discussed in Chapter 1 as it is a major heading. Clamping force is expressed in kN (kilonewtons) by European manufacturers and non-European manufacturers supplying the European market. Some Japanese manufacturers, however, continue to use tonnes, but caution has to be exercised when the term 'tons' (note the difference in spelling) is used. This is often the case with other Far Eastern manufacturers and will not refer to Imperial but US or 'short' tons. The US ton is about 10 per cent less than the Imperial ton, which in this context is practically the same as a Metric tonne (1 Imp ton = 1.12 US tons = 1.02 Metric tonne).

A 300 tonne machine in European terms is equivalent to a 330 US ton machine and conversely a machine described as a 300 tonner in a Far Eastern maker's brochure would be comparable with a European machine of about 270 tonnes or 2,700 kN. The difference is therefore considerable and must be taken into account. A prospective buyer should check the point, if it arises, with the manufacturer concerned or his agent. If 'tonnes' appears and it is certain that it means Metric tonnes, 1 tonne is almost exactly the same as 10 kN. A 100 tonne machine, for example, has a clamping force of 1,000 kN.

▲ Locking force.

At least one manufacturer also gives a figure for 'Locking force', making a distinction between this and 'Clamping force', albeit a fine one. Most makers omit a figure for this as the clamping force is of prime significance for the purposes of machine selection.

Locking force refers only to machines of the toggle and not the direct hydraulic type. With a toggle machine, the mould height adjustment is set so that the desired clamping force is exerted against the mould when this is closed and the toggle system is fully extended. In this condition, the mould is thus clamped with a force of, say, 1,200 kN. Then injection begins and the pressure of this acts against the frame formed by the extended toggle system. Because this frame is rigid, the tiebars are in turn caused to stretch, and the force exerted by stretching the tiebars is termed the 'locking force'.

With a direct hydraulic machine, the tendency during injection for the mould to open is countered by the hydraulic ram, behind which is a high pressure volume of oil in a cylinder, in effect acting like a powerful spring. Whereas the tiebars stretch with a toggle machine, the high pressure oil takes the load in a hydraulic machine.

Where stated in a specification table for a toggle machine, the figure for the locking force is usually 10 per cent higher than that for the clamping force. A machine with 1,200 kN clamping force will be shown as having a locking force of 1,320 kN. However, the clamping force figure remains the more important of the two for the purpose of selecting a machine. The supplementary effect of the locking force is determined in the design of the machine and if the manufacturer states its kN in the specification it is chiefly for the sake of reassuring the potential customer.

If a machine with a certain clamping force is selected to suit a particular mould, it can be taken for granted that the locking force will be sufficient, even if it is not stated in the specification. If the mould does flash open in the course of a practical trial, a miscalculation has been made in regard to the amount of clamping force required and a machine with greater clamping force must be selected. It is necessary to point out that certain manufacturers quote only locking force and not clamping force. This needs to be checked in relation to the clamping force required because the clamping force of the machine will be *less* than the locking force quoted. A miscalculation in matching the clamping force of the machine to requirements might be made on account of this possible confusion.

▲ Overall platen size and Distance between tiebars.

These factors have already been dealt with in Chapter 1 but an important point of detail concerns what is known as the 'mould spigot' or 'registration ring'and the size of hole allowed for it in the fixed platen of the machine.

The spigot is a raised circular feature on the backplate of the fixed mould half. This is of a certain diameter and intended to fit into a circular hole of the same diameter in the fixed platen to ensure that the mould is accurately located in the centre. The hole in the platen must be large enough for the spigot and not too small. If it is oversize, the unwanted space can be taken up by a spacer ring but there will be difficulties if the spigot is too big.

A manufacturer will supply a new machine with a hole of suitable diameter, but the point needs to be checked if one is offered a stock machine.

A table of machine specifications normally includes small platen drawings and the standard diameter of the hole allowed in the platen for mould registration will be shown.

(Left) The fixed platen of a machine has a hole in the centre through which the nozzle of the barrel of the injection unit protrudes for a short distance. When the mould is in position, the nozzle aligns and is in contact with the sprue bush, the tube for the entry of material into the mould. The central hole in the platen is of a certain diameter so that the spigot, or raised circular feature, on the mould can be accurately located in it. This ensures that the mould is properly positioned. There is also provision for adjustment on the injection carriage to enable the nozzle to be accurately centred.

(Right) This picture shows the moving platen. The hydraulic ejector behind it operates through a central hole. The ejection mechanism of the mould is connected to the hydraulic ejector and is actuated for demoulding, that is ejecting the mouldings from the mould. This machine also has four additional ejector connections around the central point.

▲ Tiebar diameter.

A few manufacturers quote the tiebar diameters for their machines. Tiebar diameter determines clamping force on a toggle machine. The greater the diameter and hence the cross sectional area of the tiebar, the greater the clamping force that can be applied. A tiebar will stretch by a few thousandths of a mm when full clamping force is exerted and a tiebar of insufficient diameter will break. (This can also happen if the clamping unit becomes out of alignment, which can happen after a period of service through poor attention and maintenance.) The grade of steel used for the tiebars has also to be taken into account as the better the quality of the steel the greater the stretching that it can withstand, but the prospective buyer is unlikely to want such details and most manufacturers omit them.

▲ Minimum and maximum mould height, Maximum mould opening stroke.

These have been discussed in Chapter 1.

▲ Maximum daylight is:-

(a) with toggle machines the maximum machine mould height plus the total opening stroke available,

and

(b) with direct hydraulic machines the minimum mould height plus the total opening stroke.

This dimension gives some indication of the capacity of the clamping unit of the machine but it is superfluous to consider it once the machine mould height and opening stroke have been considered individually, although this may be done as a cross check.

Some machine manufacturers refer to the largest imaginary rectangular box that can be mounted in between the platens of their machines. The size of this box is the width x length of the longest mould that can be mounted on the platens x the maximum machine daylight. This gives an indication of the capacity of the mould space of their machines but it must be further analysed in terms of machine mould height and stroke to ensure that each of these is acceptable.

▲ Hydraulic ejector stroke.

Early injection moulding machines had a simple mechanical means for ejecting the moulded parts from the mould cavities. This took the form of a bar, or plate, behind the moving platen which struck a pair of adjustable stops when the platen retracted. The effect was to push the bar forward relative to the platen and for the ejector pins in the mould cavities to move forward and push the parts out. If you look at an injection moulding, you may find small circular marks on the surface. These are the witness marks, or slight blemishes made by the ends of the ejector pins as they pressed against the moulding to push it out.

Mechanical ejection was a rudimentary system that has now been replaced on all but small inexpensive machines by hydraulic ejection. There is a special hydraulic cylinder mounted in the centre of the rear of the platen and its piston moves forwards to actuate the ejector pins. All electric machines have an electrically operated ejection system. The ejector can make not one but several strokes if needed to eject obstinate parts. Pins may be positively pulled back before the mould closes again. With the old mechanical system this was only possible if the pins were spring loaded.

As an alternative to pins, there is an ejection mechanism incorporating what is termed a 'stripper plate'. This may be used for mouldings of a certain shape not amenable to ejection by pins. Examples are vending cups with narrow rims. The stripper plate applies ejection force to the complete circumference of the rim.

It is also possible to actuate the cylinder while the moving platen is actually in motion and reduce cycle time. This is known as 'ejection on the fly', or 'parallel ejection', meaning that it is performed in parallel with, or at the same time as, platen opening movement.

Where this heading is given in a table, it is of relevance if the part concerned is exceptionally long and therefore requires an unusually long ejection stroke. The manufacturer will normally provide an ejection stroke of sufficient length for most requirements, but if a longer stroke than standard and as shown in the table is required, the matter may be further discussed.

When a mould is mounted in a machine, its ejection system has to be connected to the hydraulic ejector cylinder. A point to check is whether the machine has automatic means for this as standard. If not, connection and disconnection have to be performed manually. Proprietary automatic connection devices are available and can be fitted to most types of machine.

It is important for a buyer to ensure that the machine has suitable ejection arrangements if it is to be purchased for an existing mould or moulds. (See also 'Mould fixing hole pattern' page 81.)

▲ Ejection force and Retraction force.

Where 'Hydraulic ejector stroke' is stated in a table, the manufacturer may also give the ejection force and retraction force. The ejection force is the amount of force that the piston in the hydraulic ejection cylinder applies. A formula exists for the estimation of ejection force required for a moulding but in its absence whether a machine is adequate for a particular moulding is a matter that the prospective buyer may decide in the light of his or the machine supplier's experience.

Likewise with retraction force, the force with which the ejector is pulled back. It is possible for the pins to stick to a moulding, particularly if it has not fully cooled and is still soft, so that they will not come back easily. One manufacturer quotes an ejection force of 41 kN for ejection strokes of 125 and 150 mm on his 500,

800 and 1,000 kN machines and a retraction force of 15 kN, so that retraction force is about a third of ejection force. It should be possible to control the speed and force of the ejector so that witness mark blemishes may be made less noticeable.

We now come to the injection unit, although a few manufacturers put the data for this before the clamping unit.

▲ Injection unit rating.

As in the clamping unit section, this line for the rating provides convenient subheadings for the data concerning the various injection units and reflects the Euromap figure. For any given size of machine in terms of clamping force, a manufacturer will probably offer a choice of two or more injection units with different ratings or capacities. The table is subdivided into columns each of which refers to a particular injection unit rating. For example, a 1,000 kN machine may offer a choice of two injection units rated at 220 and 445 or perhaps three rated at 320, 440 and 610. These figures will be shown at the tops of the columns of specifications for the injection units.

A machinery manufacturer offers a choice of injection unit ratings in order that the customer can select the optimum combination of kN clamping force and injection capacity for his requirements. Within any injection unit rating there will furthermore be a selection of two or three screw sizes, and the table will again be subdivided for the respective details. The next line under the sub heading for the injection unit rating will show the screw sizes available for it.

The larger the diameter of the screw, the greater the volume of the shot that can be injected and the figure is shown in cm^3 further down the column. It may be titled 'Maximum shot volume', 'Injection volume' or 'Cylinder head volume'.

When a manufacturer offers an injection unit with larger cylinders and a higher rating, he usually also provides a longer stroke to give greater shot capacity in terms of volume and weight of material. For instance, a 320 rated injection unit will provide a screw stroke of 175 mm but a 440 unit will have 9 mm extra stroke at 184 mm. In turn a 610 rated unit for the same machine gives another 19 mm stroke at 203 mm. This gives the buyer a good range of choice for a machine of any kN clamping force. In addition, as mentioned, there is normally also a choice of screw sizes for any injection unit of a certain rating. Taking the 445 unit, the manufacturer offers a 40 and a 45 mm diameter screw, and for the 610 there is a choice of 40, 45 and 50 mm. But bear in mind that the 40 and 45 mm screws in the 610 unit will have greater shot volumes because longer strokes are provided. These can be seen from the tabulated figures.

▲ Screw diameter mm.

The diameter of a screw is used to define it and its associated barrel. The more important practical figure is the *shot volume* of the screw/barrel assembly, which is determined by screw diameter and injection stroke. The screw diameter defines the individual screw/barrel unit under the injection unit rating subheading, so that one might refer, for example, to the 40 mm screw of a 430 rated injection unit. For any one injection unit rating there is a choice of screw diameters. In the table, the various screw diameters available are listed and under these are the shot volumes and maximum injection pressures relative to the diameters.

Although they do not often appear in manufacturers' specification tables, the terms A, B and C are historically applied to the selection of screws available for an injection unit of a specific rating. The A, B and C screws are in descending order of maximum injection pressure available and ascending order of diameter. If an injection unit of a certain rating offers a 30, 35 and a 40 mm screw, for example, these are called the A, B and C screws respectively. Alternative terms are high, medium and low pressure screws.

There is also sometimes an H screw, which is a term for an ultra high pressure screw smaller in diameter than an A screw and possibly also longer in L/D ratio (see below). Injection pressures have markedly increased as machines have been developed over the years and the injection pressures available with the A screws of modern machines are even greater than the H screws of older models. The term H screw is therefore becoming moribund as today's A screws meet the great majority of high pressure requirements.

Generally an A screw is one with a maximum injection pressure of about 2,000 to 2,200 bar, a B screw 1,600 to 1,800 bar and a C screw 1,400 to 1,600, but for heavy machines these figures may be about 200 to 300 bar less. This is because thin wall sections are not normally featured on heavy mouldings and an injection pressure of 2,200 bar would hardly be required. The A, B and C screws of a certain 9,500 kN machine have respective maximum injection pressures of 1,850, 1,529 and 1,285 bar for shot weights of about 3.7, 4.5 and 5.4 kg.

▲ Length/diameter (L/D) ratio.

The length/diameter (L/D) ratio of the screw may be given in a table of specifications. This is an indication of screw performance but should be checked by a practical trial. Modern injection moulding machine screws usually have an L/D ratio of at least 20:1. A screw of 30 mm diameter and with a 20 L/D ratio (sometimes simply expressed as a '20D' screw) will be 600 mm long and this gives an indication of the amount of plasticising treatment, or 'kneading' as it is appropriately described, imparted to the material. A 15:1 L/D screw of similar diameter will not plasticise the material so thoroughly, but it is less expensive and may be sufficiently effective with less demanding materials. Some manufacturers offer 25D screws either as standard or optionally for improved melt homogeneity and colour dispersion.

Where there is a choice of screws of different diameters, it should be checked that the L/D ratio quoted applies to all of them and not just one. For example, if an L/D ratio of 20:1 is quoted and a certain injection unit offers a choice of screw diameters of 25, 35 and 40 mm, it may be that only the 25 mm has that L/D ratio.

This would mean that the others would have smaller ratios and their performance would not be comparable. All screws would be of the same length, but the L/D ratios of the larger diameter ones would be lower. The 25 mm screw would be 500 mm long and have an L/D ratio of 20:1, but the 40 mm screw would have an L/D ratio of 12.5:1 for the same length. Leading manufacturers apply the quoted ratio to all screws irrespective of diameter and this means that the larger diameter screws are longer than the smaller. The machines are built to accommodate screws and barrels of different length so that changeovers can be made without difficulty.

▲ Injection stroke.

This figure, like screw diameter, means little by itself but together the two determine the shot volume, which is of prime interest.

▲ Maximum injection pressure, shot volume and shot weight.

These are important basic factors in machine selection and have been dealt with above.

▲ Rate of injection.

Some moulders base their choice of machine primarily upon the rate of injection, a critical factor if a fast cycle time is required. Also, some kinds of moulding have to be injected extremely quickly. The rate of injection is a theoretical figure for the maximum amount of material that can be injected into the mould in terms of grams per second and is calculated on the basis of the shot volume and the speed at which the screw moves forward to displace it. It is theoretical because the assumption on which it is based is unreal, that is that injection is not into an actual mould but into free air and completely unimpeded. The reality is different as the material has to be forced into a mould which may have narrow sections and small diameter holes, or 'gates', into the cavities. These will slow it down considerably. One manufacturer makes it clear in his tables that the figures are based upon injection into free air but others do not and misunderstandings on this point often arise. It is not possible to take the shot volume for the moulding and divide this by the rate of injection to arrive at an exact figure for the time needed to fill the mould. Only an approximation can be obtained and this should be validated during a mould trial.

The actual rate of injection that can be attained for a particular mould can be assessed only practically, but the figure for rate of injection given in a specification does provide a basis for comparison between different models of machine. The more powerful the machine, the higher the rate of injection. The figure for rate of injection is in any case probably useful only where high filling rates are involved. Most machines have adequate rates of injection for the majority of mouldings and high filling rates are needed only for certain kinds of moulding such as disposable cups and cutlery, flower pots and other mass volume products.

▲ Maximum injection speed.

Injection speed, with shot volume, determines the rate of injection, as discussed in the preceding paragraphs. It should be understood that the figure given for maximum injection speed, in terms of mm per second, is not the speed at which the screw moves for the whole of its injection stroke but the peak maximum speed attained for only part of it. If, for example, maximum injection speed is given as 1,000 mm per second, it means that this speed is attained, or may be attained, after acceleration from a standing start.

Most manufacturers do not quote maximum injection speed, although many give the rate of injection, which is the figure that matters. Those who do give maximum injection speed include the few who make exceptionally powerful machines for mass volume products. The maximum injection speed quoted for these machines is usually 1,000 mm per second. For other machines up to about 3,000 kN the rate may be up to 400 mm per second and larger models 250 mm per second, which is adequate for most work.

High injection speeds are often required where a hot runner system is used and directs materials through its channels to a large number of cavities. Where this is the case, the volume of material in the channels is considerable and its viscoelasticity, or compressibility, has to be taken into account. A high injection speed may be required to overcome the compressibility of the material in the channels quickly before it will be sufficiently pressurised to flow through the small pinhole sized gates and into the cavities. A machine with poor power and low injection speed will noticeably show up its weakness in such circumstances and may represent a bad investment. Its output per hour will be lower than for a higher powered machine because it will need more time to pressurise the material and fill the cavities.

With injection strokes of extremely short time duration, the maximum injection speed stated and set on the machine may not even be approached (see also 'Closed loop injection control' page 96).

▲ Plasticising rate.

If fast cycling work is intended, the plasticising rate, sometimes also called the 'recovery rate', should also be taken into account. This is the maximum rate at which material can be plasticised by the screw and is usually quoted in terms of grams per second and in respect of PS (polystyrene). If another kind of material is involved, the figure quoted can be no more than a guide because some materials, PS among them, can be plasticised more quickly than others.

Material is plasticised as the screw rotates. Solid granules of plastics fall in at what is called the 'back' of the screw, that is at the hopper end, and are melted by virtue of the screw's rotational action. The molten material passes to the front of the screw (at the nozzle end) and when a sufficient amount - the shot volume - is accumulated, it is injected into the mould. Plasticising for one shot takes place after the previous shot has been injected into the mould and while the moulding is cooling and, in most instances, there is adequate time during cooling for the next shot to be plasticised by the rotation of the screw.

With fast cycling moulding, however, cooling times are short and the time for plasticising the next shot is limited. It is therefore in respect of fast cycling moulding that the plasticising rate of the machine is critical and sometimes it is necessary to prolong cooling time beyond what is strictly necessary simply to allow sufficient recovery time. The economics of the process are therefore prejudiced as the machine is not attaining its potential maximum output rate per hour.

The plasticising rate quoted for the machine has, in such instances, to be considered in relation to the cooling time needed for the moulding. This is dependent upon type of material and wall thickness. Manufacturers of materials provide tables of figures which assist in the calculation of cooling times and it is not difficult to reckon these.

For purposes of explanation, we will assume that the moulding is in PS, as is used for the figure for the plasticising rate in a machine table. A plasticising rate of 20 grams per second may be quoted for a particular injection unit and screw size. Therefore, theoretically at least, a cooling time of, say, 4 seconds should allow 80 grams of material to be plasticised. If the weight of the moulding is less than 80 grams, cooling time need not be extended, but cooling time has to be prolonged if it is greater. Cooling time has to be long enough for recovery, although on some more expensive types of machine there are ways of controlling the cycle sequence so that little or none of the available cooling time is spent without the screw plasticising the next shot. The screw can continue to rotate while the mould is being opened and closed.

The machine manufacturer should be able to advise on plasticising rates for other materials. Examination of the figure for the plasticising rate in a table, even if it does relate to PS, should, however, at least alert the prospective buyer to the possibility of a problem in this respect. A practical trial in the material concerned is the only certain way to resolve it. Some manufacturers offer the options of higher than standard pump outputs so that greater plasticising capacity can be provided. In the example quoted, a more powerful pump option raises the plasticising rate from 20 to 24 grams per second and this could be sufficient to avoid an unnecessary extension of cooling time.

One manufacturer quotes plasticising rate in terms of cm^3 per screw revolution, which is probably better than grams per second as it enables the screw rpm to be taken into account, but the length of time for plasticising at a given rate of screw revolutions per second will need to adjusted according to type of material. Some materials tend to scorch if the peripheral speed of the screw is too high (see next paragraph).

▲ Screw rotation speed.

This is sometimes quoted and means the maximum rpm (revolutions per minute) that can be attained. On the machine, screw speed is variable up to the maximum stated, although with some hydraulic screw drives low speeds may be erratic and unusable. The faster that a screw can be rotated, the greater the amount of material that can be plasticised in a given time, but there is an upper limit beyond which it cannot be taken. The crucial factor is not rpm but the peripheral speed of the screw, that is the speed in mm per second at which any point on the outer edge of the flight is travelling. The edge of the screw revolves at a certain speed but the barrel within which it is turning is static. Material in the flights of the screw between it and the barrel surface is therefore subject to a greater amount of kneading the higher the rpm.

This is technically referred to as 'shear': the molecules of the material are sheared, or shredded and torn apart, and this treatment raises its temperature. There is a limit to which the material will withstand this treatment before so much heat is built up that it scorches, that is locally burns and degrades. Screw rotation speed can therefore be taken only up to a certain limit before scorching is liable to occur. Some materials can withstand high screw speeds better than others. An approximate guide is that commonly used materials such as those used to make packaging and household articles can withstand a peripheral (also termed a 'circumferential') screw speed of 1,000 to 1,200 mm per second. It is in fact the ability of materials of this kind to accept high screw speeds that makes them suitable for fast-cycling production of high volume mouldings in the first place. Materials which are more thermally sensitive, as used to make precision mouldings for technical applications, may withstand peripheral speeds of only 300 or even as little as 100 mm per second.

It also has to be borne in mind that the diameter of the screw is a critical factor: at a given rate of rpm, the peripheral speed of a large diameter screw is greater than for a small diameter screw. Also, a material itself may withstand a high peripheral screw speed but certain additives it contains might not. Some pigments and flame retardant additives are sensitive to shear and therefore impose a limit to screw speed.

▲ Maximum screw torque.

Maximum screw torque is quoted by a few manufacturers and is an indication of the power provided by the machine pump and ultimately the plasticising or recovery rate. A high plasticising rate can be achieved only if the torque available is sufficient, and some materials require more torque than others.

Torque is defined in the dictionary as a 'twisting moment that tends to produce angular acceleration'. The term comes from the Latin word for a necklace and can be imagined as the power to turn something, in this case the screw of an injection moulding machine. Imagine you are driving a car up a steep hill. If the engine has poor torque, or power to turn the driving wheels, your speed will rapidly fall and you will have to change down gear. Another car with more torque at the same speed will enable you to keep going up the hill without changing gear. Some materials need to be plasticised at relatively low screw rpm because of sensitivity to shear. This sometimes causes problems as the hydraulic screw drive motor runs jerkily at low rpm. If this problem arises, a practical trial will show whether the maximum screw torque quoted in a specification table is in fact available at low speeds. If not, most motor designs allow an alternative interchangeable adapter to be fitted and a machine can be supplied with either high speed/low torque or low speed/high torque for the screw drive. Some specification tables show data for both high and low torque drive options.

▲ Nozzle contact force.

This is another statistic which really ought to be taken as read. The buyer has a right to expect that the nozzle contact force is adequate, but some manufacturers quote it and others don't. The nozzle is the short tube at the front end of the barrel and is brought forward and held in contact with the sprue bush, another short tube which is contained in the mould and provides the passage for the entry of molten material into the mould. The nozzle contact force is the 'push' applied behind the nozzle to ensure that it stays in contact with the sprue bush when material is being injected through it at high pressure. If nozzle contact force is inadequate, material will squirt out in all directions because the nozzle will be lifted away from the mould.

A properly designed machine has one or a pair of substantial hydraulic cylinders which impart the necessary force to the injection carriage, of which the screw/barrel assembly, with the nozzle, is part. It is unlikely that it will be insufficient.

The nozzle should be correctly aligned with the sprue bush and there is usually a means for adjustment of this. Incorrect alignment causes loss of injection pressure as the hole in the end of the nozzle and that at the start of the sprue bush will not line up perfectly and in effect the flow path through is restricted. A well made machine will maintain good nozzle alignment throughout its life, but one less so may have inadequate mechanical support for its injection unit on the base frame and over a period of years will tend not to maintain accurate nozzle alignment without frequent attention and readjustment. This is a serious fundamental weakness that will affect its profitability.

▲ Carriage stroke.

The same cylinder or cylinders that provide nozzle contact force also enable the injection carriage to be retracted away from the fixed platen over a certain distance known as the carriage stroke. Carriage retraction may be necessary for attention to the nozzle or a heater band, or for the screw/barrel assembly to be changed. Many machines also allow the injection unit to be swung out to the front to improve access to these parts. Swinging out the barrel can be done simply manually, but hydraulic power for this purpose may be a standard or optional feature.

An important point concerning carriage stroke is that sufficient space must be allowed for it when the machine is placed in position on the factory floor. A diagram in the machine brochure or manual will show the dimension concerned. Don't place the machine with the screw drive motor only a few inches away from a wall.

▲ Hopper capacity.

This heading, where it appears, is a hangover from the days when material was loaded manually into the hopper from sacks. This practice has become redundant except for small machines because loading is generally performed automatically either by individual pieces of hopper loading equipment or from a centralised materials handling system. The hopper capacity figure enabled the moulder to work out how often he needed to replenish the machine according to the amount of material it consumed per hour. As automatic loaders have become de rigueur, some manufacturers offer a hopper only as an optional extra and no longer as standard equipment.

▲ Oil tank capacity.

Although this is a piece of information more relevant to machine installation than selection, it will have a relation to running costs if the owner makes regular oil changes, as recommended. Machine design has progressively tended to reduce the amount of oil needed to run a machine and comparisons may be made between different types.

▲ Pump kW.

Buried away towards the bottom of the table, pump kW is nevertheless a fairly important parameter although its significance has to be interpreted carefully. Whilst it is broadly the case that the more powerful the pump, the more efficiently and faster the machine will perform, the kW figure can be excessive. Some manufacturers will also claim that the efficiency of their hydraulic systems makes it possible to incorporate a less powerful pump than their competitors, so saving energy. It is a matter of the pump kW being correctly matched to requirements. An electric motor runs most economically when its maximum output matches requirements and less economically when under partial load, although modern motor control devices optimise output to some extent.

This figure is probably less helpful than others discussed above in assessing the perfomance of a machine but may be read in conjunction with them to compare machines of different makes. It may also be higher with direct hydraulic machines than toggle as a greater volume of oil has to be pumped around the clamping system for the same platen speeds to be achieved. A practical trial reveals whether the machine has adequate power for the work intended and allows comparative specific energy consumption figures, that is energy used in relation to the amount of material processed, to be obtained.

▲ Cylinder heating power.

The injection barrel, or cylinder, is fitted with a number of heater bands and there is also a smaller band around the nozzle. The cylinder heating power is the aggregate kilowattage of these heaters. Comparison between machines may show differences. The amount of heating power in relation to shot volume tends to reduce the larger the size of the machine. If long cooling times are involved in moulding certain thickwalled components, a practical trial will reveal whether the amount of heating power is inadequate, but this is unlikely.

The main disadvantage of low cylinder heating power is that warm up time, that is the time needed to bring the screw and barrel up to working temperature from a cold start, is prolonged. Heaters of low power may need about 30 minutes more warm up time to bring a machine of about 1,500 kN up to working temperature from cold than if higher wattage heaters are fitted.

▲ No. of heating zones.

The temperature along the barrel is controlled so that it increases from a minimum setting at the hopper end to maximum at the front. The length of the barrel is divided into a number of zones and the temperature of each zone is individually regulated. There are usually two, three or four zones of control plus an additional zone for the nozzle. The information against this heading will therefore read, for example, 1 + 4, meaning that there are four zones along the length of the barrel plus a nozzle zone.

The barrel zones are numbered beginning with no.1 at the hopper end, where the material enters. The series of zone numbers 1, 2 and 3 therefore runs from right to left. The zones are not numbered from left to right, as might be expected, because they follow the flow of the material. The greater the number of zones, the more smoothly the rise in temperature can be controlled as the individual steps from one zone to the next are more gradual. Machines in a higher price bracket may have more control zones than less expensive models and this point is worth checking.

▲ Total installed power.

This is another piece of installation information and means the pump kW plus the kW consumption of all the barrel and nozzle heater bands and an oil reservoir heater, if fitted. If the pump runs at full output and all the heater bands are working, their aggregate consumption is equal to the total installed power. It is unlikely that this condition is ever realised but the electrical connection for the machine installation has to allow for it.

▲ Line pressure.

Again an important piece of information almost casually included in the table. Some makers do not in fact state it. Most machines operate at a maximum hydraulic circuit pressure of 140 bar but some go up higher, even to 210 bar. The high pressure permits fast movements and better response to control signals by vital parts such as the screw and the moving platen. It also reduces the amount of hydraulic oil necessary to operate the machine.

Where the figure is omitted, it can be assumed that the machine has a line pressure of 140 bar, but the manufacturer can be consulted if there is doubt on this point. Early injection moulding machines, as a matter of interest, had 70 bar circuits. Pressures went up so that the physical size of cylinders and other hydraulic gear could be reduced. Almost all machinery manufacturers fit hydraulic equipment made by proprietary manufacturers, most prominent among whom are Bosch, Rexroth, Vickers, Danfoss and Atos, also Moog, a specialist maker of 'servo' valves for closed loop injection control. A few machinery manufacturers make some valves of their own. It is rare for a machine to have all of its hydraulic components from a single source, even although this tends to be reassuring and may simplify the holding of spares. A machine is more likely to have a mixture of makes of hydraulic equipment because machinery designers have preferences for particular valves for certain applications in the circuit. Also, ups and downs in the economic cycle sometimes create shortages of hydraulic valves and machinery manufacturers have to fit what they can get at any particular time, creating further non-uniformity.

▲ Dry cycle rate.

'Dry' in this expression means 'dummy'. The machine goes through the motions of moulding something as fast as possible but doesn't actually process any material. No cooling time is therefore involved. A moulder may wish to know how ultimately fast the machine can run on the hypothetical basis that there is no cooling period. The dry cycle rate, expressed in cycles per hour, gives this information. A cycle means one performance of the complete sequence of processing phases by the machine to make a part, or number of parts at once according to how many mould cavities there are. One cycle is made repetitively after another so that the machine constantly maintains the moulding process on a cyclical basis.

There is a Euromap formula for the sequence of movements to be performed during a dry cycle to ensure that all manufacturers give rates reckoned on comparable terms. The formula allows nozzle retraction to be included or excluded. A machine may, for example, be said to have a dry cycle rate of 1,600 per hour with nozzle retraction or 2,100 per hour without. Some mouldings are made with nozzle retraction and others not and the appropriate figure is applied. The time, as distinct from the rate, for a dry cycle might alternatively be given, for example 1.5 seconds. There are 3,600 seconds in an hour and dividing 1.5 seconds into this gives a rate of 2,400 per hour. Quoting the time instead of the rate may make it easier to calculate the number of cycles which can be made in actual operation. The cooling time is added to the dry cycle time. Assuming that screw recovery can be performed within the cooling time, the number of cycles per hour which can be performed can be calculated in this manner.

However, running at the maximum possible cycle rate means that the machine is at its utmost limit. Whilst this may be theoretically possible, it is the smoothness, stability and precision actually maintained by the machine during rapid cycling that determine how fast it can perform satisfactorily. These factors are determined by quality of design and build. A poor quality machine will not last for long if cycled at its ultimate rate and the standard of its output will soon decline. Unless build quality is carefully assessed, this figure can be deceptive.

▲ Net weight.

Although this figure is chiefly given so that transport and shipping costs can be calculated, or perhaps floor loading checked, it is in fact a good guide to build quality and the potential longevity of the machine. A 1,000 kN machine of basic to medium quality might weigh 4,000 to 6,000 kg but an expensive high quality model nearly 10,000 kg. Different makers' specifications may be examined to draw comparisons of this kind. Prices of machines tend to correlate to weight to some extent.

Manufacturers do not deliberately set out to make their machines heavy, if only because the use of more metal puts up the price, but some makers will build down to a price and use as little metal as possible. At the other end of the scale, a high quality machine will be heavy partly because it is fitted with more hydraulic valves and manifolds for refined control but chiefly because it is basically designed on a more substantial scale.

The platens and clamping system, for example, will be heavier than on other machines to resist deformation under the stresses of clamping and injection and the base frame will be sturdier. When the moving platen closes, it is decelerated sharply, that is positively 'braked', before the mould faces meet; similarly when it opens it is decelerated before it reaches the end of its stroke. Both braking phases involve a considerable dissipation of energy, remembering that not only is the platen a heavy part in itself but so is the mould half bolted to it. The dissipation of energy at the end of each stroke is absorbed by the mass of the machine and the more metal this comprises the longer the life of the machine is likely to be. The figure for the weight of the machine can therefore provide an indication of life potential and be taken into account with its price.

▲ Dimensions.

The production floor of a typical moulding plant usually represents only a minor proportion of the total building area, most of which is given over to storage and handling of the parts produced, even with Just in Time production. Downstream operations such as printing and decorating are also space-consuming. One might think that the dimensions of the machines should determine the size of the production area, but in most instances machines have to fit into allotted spaces. Dimensions are therefore often critical and there are many instances of where a compact machine has been preferred to a bigger type of equivalent performance. Where this is so, these figures need to considered at an early stage of the selection process. A specification table is usually accompanied by diagrams showing the machine in three elevations. The footprint, or floor space occupied, is the bird's eye view. If the control cabinet is separate, this must not be overlooked. Space must also be allowed around the machine for access and any dismantling of the machine in situ, such as tiebar replacement, that might be undertaken.

If the machine is to take the same place as an old one, it will probably have larger platens and a longer daylight, but more up to date base frame design and tidier looping of the hydraulic hoses may result in the footprint being similar. Make sure that the full retraction of the injection carriage is allowed for. This will be shown in the plan diagram.

Overhead clearance should also not be ignored. Whilst the machine itself may have plenty of clearance, a takeout robot of the type with a tall vertical mast mounted on the top of the fixed platen may interfere with the roof purlins or a travelling crane. This problem is likely to arise only with large machines of 10,000 kN or more. Some manufacturers in this class offer models of special low design. Many big machines are installed with their base frames sunk in pits and their centreline at convenient height relative to floor level.

The centreline, or axis, of a machine is a line drawn through the screw and the centre of the clamping unit. It should be shown on the drawings of the machine and is an important detail. Generally, the lower the centreline, the easier it is to work on the machine. The height of the centreline from the floor of a 1,000 kN machine may be about 1,300 mm, a 1,500 kN 1,400 mm and a 5,000 kN 1,600 mm. On a machine of about 20,000 kN this dimension might be 2,200 mm and on a mammoth 60,000 kN nearly 3,000 mm, which would probably make excavation necessary. The machine axis is the reference or datum line used for relating the position of any special equipment such as a takeout robot.

chapter 3

Optional extras, alternative equipment and ancillaries.

Manufacturers compete with each other in respect of the range of standard features they include in their machines as against offering these as optional extras at additional cost. A standard machine specification should be checked to see what features are included or excluded. A higher price for one machine compared with another may be seen to be justified when the more comprehensive range of its standard features is taken into account.
The following items are taken from the lists of optional extras and alternatives for a number of makes, but some options may be standard on certain machines.

▲ Injection unit.

Insulated hopper/heated hopper.

Materials are often predried and preheated in a centralised drying installation or a special mobile drying unit before being processed on a machine and an insulated hopper helps to keep them dry once they have been fed into it and until they fall into the screw/barrel unit for processing. A heated hopper will do the same but more positively. More powerful types of heated hopper may be used instead of a centralised installation or mobile unit to dry damp material.

Hopper magnet.

This is a grating of bar magnets placed at the bottom of the hopper. As material flows through it, any magnetic metal inclusions will be pulled out, which is obviously desirable. Many metal particles which are non magnetic, however, are likely to contaminate material, particularly if materials handling methods are open, that is not sealed off from the ambient environment. When scrap material is regranulated, it may be contaminated by small fragments from the chopping blades of the granulator, which could be of a non magnetic steel alloy. The use of an electronic separator is advisablewhere there is this risk. Such fragments may lodge between the check ring and its seat (page92) and prevent proper shut off of melt during injection.

Temperature control of feed zone.

This feature is sometimes called 'Closed loop control of throat temperature'. When the granules of plastics fall into the barrel at the feed zone, they must be prevented from melting immediately at that point because the intake of material will be impeded if this happens. A blockage will be formed by a build up of molten material. The machine therefore has as standard a channel for a flow of cooling water around the feed throat, that is the orifice leading down from the hopper into the barrel. Normally the flow of water is uncontrolled and the temperature at the throat may vary as a result. Some machine manufacturers recommend that a water temperature controller be fitted at this point so that a consistent temperature at the feed zone can be maintained. By stabilising temperature in this manner, the subsequent input of heat into the material as it progresses along the barrel is more consistent cycle to cycle.

Electric screw drive.

Whilst there are a few all electric injection moulding machines on the market (which by definition include electric screw drive), most machines have hydraulic drive systems and a hydraulic motor is used to rotate the screw, an action which plasticises the solid granules drawn into the barrel from the hopper. It should be understood that the screw drive motor only rotates the screw: it does not push it forward to inject the material into the mould. This is a linear action, that is a straight push, performed by the injection cylinder or cylinders.

A hydraulic motor has a number of cylinders and pistons arranged either radially or axially. Oil is forced into the cylinders at high pressure from the machine's main pump and this causes the pistons to crank the screw round. At one time many makers equipped their machines with electric screw drive motors in preference to hydraulic, but electric motor design has advanced enormously in recent years and the type of motor offered today as an alternative to hydraulic for screw drive offers a number of major advantages.

These are chiefly offered for medium-sized machines used for fast cycling work involving a high hourly throughput of material because they enable high plasticising or recovery rates to be attained. Larger sizes of machine also benefit from electric screw drive because oil reservoir capacity can be reduced and with it the sheer physical size of the machine. Electric screw drive also requires considerably less energy, but this option costs more than standard hydraulic drive and whether it is advantageous in any particular instance is a matter for cost evaluation.

Closed loop injection control.

See page 96.

Mixing head, static mixer, Cavity Transfer Mixer.

Plastics moulding material may be obtained either ready coloured (colour compounded) or in a natural uncoloured state to which a small dose of pigment is added before it is run in the machine. The most popular means for pigmentation is known as masterbatch, a dense colourant in granular form normally added in the proportion of 1 to 4 per cent. The screw by itself might not be capable of dispersing the colour through the material without evident streaks in the mouldings and machine manufacturers offer special screws with an additional mixing head at the front so that even dispersion of colourant can be achieved. These also enable the amount of colourant used, which is expensive, to be reduced.

The normal flight extends almost to the end of the screw but the final section has a special pattern to stir the melt vigorously and thoroughly disperse the colourant. Mixing head designs include knurling, rows of pins and what is known as a Maddock section which reverses the flow of material and then turns it forwards again. The machine manufacturer will offer the most suitable type for the application. The extra cost of a mixing head may be compared against the savings likely to be gained from using less masterbatch.

It is also possible to fit a static mixer on the nozzle through which the melt will flow when being injected into the mould. Inside a static mixer is a series of baffles which create turbulence in the flow and enhance colour dispersion.

Another piece of equipment to improve dispersion is the Cavity Transfer Mixer, a manufacturing licence for which is held by certain machine manufacturers. The melt is transferred from one to another in a series of bowl-shaped cavities and thoroughly mixed in the process.

Vented barrel.

Some materials are more hygroscopic than others, that is they absorb more moisture from the air. Unless this moisture is got rid of, it will turn to steam in the moulding process and cause blemishes on the moulding as well as other problems. It is possible to predry the material in a proprietary drying unit. This provides for a stream of warm dry air to be passed through the material contained in a special hopper to extract the moisture. Some moulders, however, find this is expensive in terms of energy costs and prefer to use a vented barrel. Although this does not always dry as thoroughly as a warm air drying unit, the results are good enough in many applications.

A vented barrel has two stages. In the first the material is sufficiently plasticised by the screw so that any moisture content is vapourised. At the point along the length of the barrel where this occurs there is a hole in the top for the escape of the vapour, and it may be drawn out by the factory fume extraction system. The material continues its plasticising treatment from the point of the hole through the second stage and is finally injected into the mould.

Drying material by venting uses no special energy as does a drier, although more energy is required to rotate the screw as this is considerably longer than that of a non-vented barrel. The rate of material throughput with a vented barrel of a certain size can be up to 40 per cent less than for a non-vented, and this should be checked. A vented barrel is longer than a normal type and may mean that the machine is also greater in length and will take up more floor space.

Vented barrels are mainly used for ABS material. This is widely used for moulding parts of telephones and other electrical appliances because of its attractive, glossy appearance. One can visit various ABS moulders and see either driers or vented barrels in use. The choice of either largely depends upon personal preference and experience. Vented barrels may also be used with PVC, for which their purpose is to remove undesirable volatiles. Hydrochloric acid is formed during PVC processing by the combination of moisture with the chlorine content of the material.

Some vented barrels may be of poor design and material will leak out of the vent hole. A practical trial may be made to see if it does. The true situation is that vented barrels have declined in popularity and many machinery manufacturers are in fact reluctant to offer them, particularly as methods of drying material have been greatly improved. The difference between a<u>b</u>sorption and a<u>d</u>sorption is relevant here. Absorption is when moisture soaks into the material, as with a sponge. Adsorption is when it adheres to the surface of the granules. A quantity of sand, for example, will hold moisture but this is by adsorption, not absorption. Materials which absorb moisture need longer drying than those which only adsorb it. ABS adsorbs moisture and some manufacturers of this material even state that it needs no predrying, even if practical moulders disregard this advice. Nylon, by comparison, absorbs moisture. It often needs to be predried and, after moulding, soaked in water to restore its natural water content of about 1 per cent.

Barrier screws.

Some manufacturers offer what are called 'barrier screws' for either general purpose or material-specific applications. A barrier screw has not one but two threads, and one thread is not as high as the other. There is also a difference in the pitch of the threads, that is the distance from flight to flight. The effect of a barrier screw is to provide more thorough plasticising and a homogeneous melt can be obtained at a lower temperature than with a conventional screw. The difference with polypropylene can be 250 degrees C with a standard single thread screw and 220 degrees C with a barrier screw.

The great advantage is that cooling time needed is shorter and more mouldings per hour can be produced. Colour dispersion is also improved. Barrier screws are therefore worthy of investigation if available.

Bimetallic barrels, armoured screws.

Screws and barrels wear quickly when processing materials containing glass or other ceramic fibres or fillers and these options are available as alternatives to the standard parts. Polypropylene and nylon materials are commonly filled or reinforced in this way to make the mouldings produced in them stronger and more rigid. High rates of wear are also likely to be incurred with fast cycling machines when unfilled commodity plastics, that is commonly used materials which contain no glass, talc or fibres, are processed. Some of these are fairly abrasive just as they are in their plain state and a high rate of throughput will cause corresponding wear.

A bimetallic barrel has an inner liner or sleeve made of hard high grade steel and armoured screws have extra hard steel deposited in the areas where rapid wear is likely to occur. There are various proprietary treatments for coating screws and barrels (as well as moulds) with wear resistant alloys, and the ion implantation process is also available.

Sliding shut off nozzle, spring loaded needle shut off nozzle, hydraulic needle shut off nozzle, hydraulic crossbolt shut off nozzle.

Injection moulding may be done either with the nozzle at the front of the barrel in continuous contact with the mould, as with a hot runner system, or with the nozzle retracted for a short distance after the mould has been filled. The nozzle resumes contact with the mould at the start of the next cycle. This retraction movement is known as 'sprue break'. The sprue may have to be physically broken by retracting the nozzle to enable the moulding to be drawn away from the fixed mould half when the mould opens. Another reason for nozzle retraction is to prevent the conduction of heat from the nozzle into the mould.

If the nozzle is retracted, it may be necessary to positively shut it by means of a valve of some kind to prevent material from leaking out from the front of the barrel. Screw rotation may be performed before or after sprue break. A nozzle shut off valve will certainly be necessary if it is done after sprue break to prevent leakage as the material will be under pressure at the front of the barrel through the effect of screw rotation.

There are various type of shut off nozzle and the selection of the most suitable may be discussed with the machinery supplier. The nozzle shut off is not to be confused with the check ring at the end of the screw (see page 92). This has a different function.

Melt decompression, screw suckback.

When sprue break is not used and the nozzle is kept forward, that is in contact with the sprue bush, it may be desirable for the screw to be positively pulled back for a short distance, according to the moulding conditions involved. This is known as melt decompression or screw suckback and is provided in addition to cancelling the holding pressure. The weight and inertia of the screw, with that of the injection piston and screw drive motor behind it (plus the friction of the various surfaces relative to each other), effectively form a backstop against which the melt will react, or push. This may not be desirable as, again bearing in mind the viscoelastic behaviour of the melt, it may be pressurised whereas it could be better for it to relax. This condition may apply in hot runner moulding. It may be advisable to check whether melt decompression is offered as standard or is an optional extra at additional cost.

Cavity pressure changeover.

When a shot of material is injected into a mould, there is a critical point at which the fast forward movement of the screw should be halted, after which the screw applies pressure at a lower level as the injected material cools and solidifies. This is called the V/P (velocity/pressure) changeover point, or changeover from first to second stage of injection pressure. It will be understood that if the V/P changeover is too soon the mould will not be completely filled, and that if it is too late the mould will be overpacked with material. In most cases the changeover is performed once 95 to 98 per cent of the shot has been injected. As the moulding cools, it contracts and the remaining 2 to 5 per cent of the shot can be entered into the mould cavity at the lower pressure to make good, or 'top up', the space resulting from the contraction of the material already in there.

Normal practice is to set the V/P changeover point according to the distance that the screw travels forward from the start of its stroke. This is called 'stroke dependent' changeover. Another method is to set changeover according to hydraulic line pressure ('line pressure dependent'). As the screw injects more material into the mould, it gets harder for it to do so as the mould fills up. The hydraulic pressure applied to move the screw forward therefore rises. A pressure sensor in the machine hydraulic system senses this rise of pressure and it can be set to trip out at a certain point and effect the V/P changeover. However, this method is not widely preferred although it is available on most machines. Most moulders consider that the line pressure sensor is too far away from the mould cavity to be sufficiently accurate or consistent.

Modern pressure sensors of the strain gauge and piezo electric types are robust units resistant to high temperatures which can be fitted into the mould cavity itself to signal when cavity pressure reaches the critical level. They are directly actuated by the pressure of the material in the cavity itself. This is cavity pressure dependent changeover and yields results which allow mouldings to be made to a high standard of dimensional accuracy, but it is more expensive than simply employing the stroke dependent method because of the cost of the sensor, or transducer. Also the machine control system requires a special interface for the sensor. The difference between the two types of cavity pressure sensor need not be gone into here: both give highly accurate performance.

Infrared (IR) melt temperature measurement.

Conventional temperature measurement on an injection moulding machine is by means of what are known as thermocouples. A natural phenomenon is that an electric current will run from one metal to another where these are dissimilar. The strength of the current increases with rise in temperature.

Thermocouples, that is short lengths of dissimilar metals joined together, are made of particular metals which in combination will produce a current which can be accurately measured by a galvanometer or voltmeter. A thermocouple may be placed on the surface of the barrel or the injection nozzle of a machine or down a special hole drilled for it. The first kind is called a surface thermocouple and the second a deep thermocouple. The purpose of both is to indicate the temperature of the melt in the centre of the barrel or nozzle.

Either kind depends upon the conduction of heat from the melt through a mass of metal, which is slow, although a deep thermocouple, being closer to the melt, will give an earlier indication than a surface thermocouple. Neither is actually measuring the temperature of the melt itself. They are said to provide a 'reference temperature' from which the actual melt temperature can be estimated.

The introduction of the infrared temperature sensor should enable the injection moulding process to be greatly advanced. An IR sensor is placed in the barrel or nozzle of the machine and has a diamond or sapphire window. Melt passes directly over the window and its IR radiation, which varies according to its temperature, is transmitted via the window to the temperature control instrument. Response to changes in melt temperature is many times faster than with thermocouple measurement and can be detected over periods measurable in milliseconds, something far beyond the capabilities of the older method.

Gas injection equipment.

Gas injection is a technique to produce mouldings which appear to be compact or solid all the way through but which in fact are hollow. It offers many advantages besides enabling weight and material to be saved. An amount of molten material ('melt') is injected into a mould cavity to fill it only by about a half or two thirds. Nitrogen gas is then injected and the effect of this is to inflate the quantity of melt like a balloon so that it completely fills the cavity. The gas is evacuated once the material has cooled and solidified and the moulding is extracted in the normal way.

Cycle times can sometimes, but not always, be reduced because less cooling time is needed than if the parts were made completely solid. More mouldings may therefore be made per hour. This sometimes may not be the case because a valve has to be opened and closed to inject the gas and evacuate it and the time needed for this may offset savings in cooling time. It is necessary to evacuate the gas because a moulding with a typical internal gas pressure of 60 bar would burst when the mould was opened.

A major advantage is that for a machine with a given size of platen the amount of clamping force in kN applied can be much lower compared with conventional moulding. Machines will probably be designed and built specifically for the gas injection process and have low clamping force relative to platen size. They should be lower in price than conventional machines.

Gas injection may also be used to assist in making large area parts such as trays. An amount of material is injected into the mould and then gas on one side. The effect is to apply pressure against the one side of the tray, usually the bottom, so that the top is pressed against its half of the mould. Clamping pressure needed may again be much less than with conventional moulding and thinwall articles produced which would otherwise be difficult to make.

Pulsed cooling.

This is a method of cooling the mould by synchronising the start of the flow of cooling water through it with the end of the injection phase, that is once the mould has been filled with melt. Equipment for pulsed cooling is linked to the machine's control system for the initiation of water flow. Cooling ends when the required amount of heat has been extracted.

Suppliers of equipment for pulsed cooling offer flush control panels which can be inbuilt by the machine manufacturer into the machine's control box so that there is no separate cabinet for it. This option may be considered if it is decided to buy a new machine and adopt the pulsed cooling technique. The conventional method of mould cooling is to control the flow of water independently of the machine cycle.

Screw back rotation lock.

This is chiefly intended for machines used for moulding thermoset materials and other machines for processing rigid PVC.

Thermoset materials such as phenolic have an extremely low viscosity in their molten state, that is they are very thin and runny. When the screw is pushed forward to inject the melt into the mould, there is a strong tendency for it to flow back over the front of the screw through the tight clearance between the screw and the wall of the barrel. When it reaches the screw flights it imparts an Archimedean action to them, that is it will tend to rotate the screw. This is undesirable and the object of providing a screw back rotation lock is simply to check this back flow of the melt. The more material that is allowed to flow back, the less that is injected into the mould and full control of the process is lost.

Rigid PVC (also termed uPVC or PVCu, the 'u' standing for 'unplasticised') is a thermoplastic material which is extremely thermally sensitive and will scorch easily if forced through narrow gaps. For this reason it is not possible to use a shut off valve at the end of a screw, as is employed for other materials, because rigid PVC would be liable to scorch as it went through during the plasticising phase. The effect of scorching is known as degradation, meaning that the quality of the material is degraded. Screws for this material therefore do not have a shut off valve but simply what is described as a 'smear head', which is just a point like the end of a lance. There is therefore nothing to check back flow on injection and the molten material, although it is not as low in viscosity as a thermoset, can flow back as far as the screw flights and cause the screw to rotate. For the same reasons as apply to thermoset moulding, it is desirable to check this. Modern PVC alloys have greater resistance to scorching than older grades, but a rotation lock will be needed if an older grade is used with a smear head.

Back flow, as explained, tends to rotate the screw and the screw drive motor will also tend to turn as a result. This will cause oil to flow out of the motor and back towards the pump, but closing a special valve to shut off this escape of oil will prevent the motor from turning and have the effect of locking the screw. On bigger machines, a disc brake of the same kind as used on a car is fitted to prevent screw rotation and back flow on injection.

Marbling equipment.

It is possible to produce injection mouldings with an appearance like marble or tortoiseshell and equipment for this can be supplied. Two or more colours are mixed together but the degree of mixing is incomplete so that the individual colours appear as swirls in the finished moulding. Cosmetic and fancy packaging is often made with marbling equipment.

Open loop nozzle control.

A thermocouple temperature sensor is dispensed with in open loop nozzle control. The nozzle is the part of the machine most exposed to risk of damage and nozzle thermocouples tend to need frequent replacement. This involves expense and downtime and many moulders choose to avoid it by running their machines with open loop control instead of the standard closed loop with a thermocouple. Open loop is also less expensive and some buyers make a small saving in the price of the machine by opting for it. However, technical components invariably require closed loop nozzle control.

There was at one time a reason to prefer simple rheostatic nozzle control, which would today be known as 'open loop', because it provided a more reliable means for regulating nozzle temperature than thermocouple control when materials such as nylon were being processed. These change suddenly with drop in temperature from the molten to the solid state. Before an improvement in the responsiveness of temperature control systems was achieved by the application of electronics, the risk of the nozzle 'freezing off', that is for the material in it to cool and solidify, could be avoided by using a simple control and setting it to a suitable level. Moulders adapted controls of the type used on kitchen cookers and the expression 'Simmerstat nozzle control', which may still persist, was originated.

Barrel heating lowered to preset value.

If a machine has to be shut down for a short time for any reason, barrel temperature may be lowered to a preset value, instead of being turned off completely, to reduce warm up time when it is restarted. This feature enables this to be done quickly without need to adjust and readjust a number of individual instruments.

Hydraulic injection unit swing out.

Many machines have a pivoted injection unit which can be swung out to the front for easy access to the nozzle or the screw shut off valve. Swinging the unit out may be done manually but this option permits it to be performed under hydraulic power to save time and effort.

Barrel insulation.

Mineral fibre material with good insulation properties and resistance to high temperatures is fitted as standard around the barrel by most European makers but offered as an option by others. Insulation conserves energy and provides a measure of safety for operators.

High temperature heater bands.

Heater bands (sometimes also called band heaters) fitted around the barrel and nozzle of the injection unit of a machine are of two main types. With the mica type, the wire element is sandwiched between two thin sheets of electrically insulating mica which in turn are sandwiched between outer plates of stainless steel. A cement adhesive is used to bond the sandwich layers together but is resistant to heat only up to a certain temperature level. The ceramic type of heater band was developed when engineering materials with higher processing temperatures were introduced. This type is made up of curved 'tiles' or sections in brick-like ceramic material and there are holes running through them in which the spiral elements are contained.

Mica heater bands are considerably less expensive than ceramic, which have to be handled carefully as they can relatively easily be broken, but nearly all European machines are fitted with ceramic as standard. There are also cast aluminium heater bands, which are robust but even more expensive. Where high temperature heater bands are mentioned in a specification, it is necessary to ask what this means. It may mean either that they are otherwise of the standard type but have a higher watt density (which means that they have a greater length of heating element) or are of a completely different type from the standard, for example ceramic instead of mica.

▲ Clamping unit.

Mould fixing hole pattern, T-slots.

Threaded holes are provided in the platens for mounting the mould halves and their pattern will normally be to Euromap standards based on Metric measurements and depicted in the diagrams accompanying the specifications, but the American SPI (Society of the Plastics Industry) standard hole pattern, based on inches, is usually available as an alternative. Moulds made in the US normally conform to the SPI standard and tools of such origin often need to be run on machines outside that country. However, it is not absolutely necessary to have an SPI hole pattern to fix US-made moulds. They can be mounted on Euromap pattern platens by means of mould clamps instead of being bolted directly into them.

There is also a hole pattern to the German DIN (Deutsche Industrie Normen) standard and a number of UK moulders adopt this.

Another alternative is the T-slot pattern. Instead of having a number of threaded holes, there are horizontal and vertical slots in the platen that allow the mould mounting bolts to be positioned where required and accommodate moulds made to either Euromap or SPI standards. T-slotted machines are popular among Italian moulders but are slow to be adopted in other countries including the UK.

The Euromap standard fixing hole pattern also provides for a central hole in the moving platen for the hydraulic ejection cylinder. The SPI standard is slightly different in that, according to size of machine, four or more holes in the moving platen are arranged in pairs at various distances from the centre. Where moulds of US origin are likely to be run in a machine, as when international companies transfer tools from one country to another for local production, a suitable ejector hole arrangement has to be provided and this is available as an option. It is not necessary for the fixing hole pattern also be to SPI standard: ejection holes to SPI standard can be provided on a platen with fixing holes to Euromap standard. The machine manufacturer can advise on this point.

Magnetic platens.

Magnetic platens are offered by some manufacturers as an alternative to traditional platens with bolt holes. Widely used in vertical presses in the sheet metalforming industry, their advantage is that they permit moulds to be quickly mounted in position and taken out again. The platens are not permanently magnetised. Their magnetism is switched on and off as required by means of an electromagnetic induction system. The magnetic field is closely limited to the area in which it is required, that is the interface between the platen and the mould half.

Quick ejector coupling.

The hydraulic ejector on the machine needs to be connected to the ejection system within the mould when the mould is fitted in the machine. This is done either manually or more quickly by means of a power operated coupling mechanism. This feature also disconnects quickly when the mould is taken out.

Hydraulic unscrewing device.

When female threaded parts such as bottle caps are produced, they are moulded around threaded male cores. This option is a special electric motor to rotate or unscrew the cores so that the caps are released for extraction from the mould.

Power operated guard door.

Access to the mould space of a machine is gained by sliding back the guard door and the next machine cycle cannot be performed until it is closed again. On small machines the door is opened and closed manually and on big machines it is power operated, either pneumatically (air) or hydraulically. Power operation on a medium sized machine may be standard or offered as an optional alternative to manual. Most injection moulding is performed with fully automatic cycling, that is the machine starts each cycle automatically, but where it is necessary for the moulding to be taken out of the mould by hand the machine has to be run what is termed 'semi automatically'. This involves opening the guard door, taking out the moulding and closing the door again. The machine can then perform the next cycle. Power operation may be desirable if opening and closing the door involves too much repetitive physical effort.

Additional cylinders for extra mould opening force.

Greater than normal mould opening force may be needed for vessel-type mouldings, for example deep pails or garbage containers. The moulding cools and contracts and in doing so tightly grips the male core of the moving half of the mould. Opening is made even more difficult if the cavity has inadvertently been overpacked with material, as may occur during initial mould setting if the injection stroke is made too long. This has the effect of jamming the two halves of the mould together extremely tightly.
On some machines it is possible to boost the hydraulic power needed to open

the mould and on others a pair of additional hydraulic cylinders may be fitted for this purpose. Opening an obstinate mould is called 'mould cracking' because a loud crack is heard when it is achieved. This is due to breaking the vacuum. The two halves of the mould are held together by the atmospheric pressure outside them because the air which was inside, in the cavity, has been expelled by the injection of the material. When they separate, the air rushes in with a loud report. A machine should normally have enough force to open a mould unless this is of unusual size and shape, but the point may be one to check during a machine trial.

Automatic clamping force control.

This option is generally available only on toggle machines of high specification. When a machine is initially set up and clamping force is adjusted to the amount required, the mould is at ambient temperature but once it has attained working temperature it will have expanded. This will have the effect of increasing the clamping force acting upon it, which may be undesirable as critical internal dimensions will be altered. Changes in ambient temperature over the day and night may also have an effect upon the expansion and contraction of the machine and mould so that clamping force will tend to deviate from the original setting.
An automatic clamping force control system employs an electronic sensor placed in a suitable position on the machine clamping unit to provide measurements over distances of a few microns from which clamping force in kN may be extrapolated. Most systems of this kind measure the extension of one of the tiebars and clamping force can be calculated from this. A means for monitoring clamping force in kN is thus available and it is a short step to linking this to the control of the machine's central mould height adjustment system so that constant clamping force is applied irrespective of temperature fluctuations. Mould height is adjusted according to any differences in tiebar extension that occur.
A variation of this system involves constant monitoring of the extension of all four tiebars to ensure that the clamping force is equally distributed over the projected area of the moulding and no one tiebar takes more strain than the others. This reduces the possibility of flashing and ensures that the machine is not subject to unduly hard treatment.
Direct hydraulic machines inherently apply consistent clamping force by the nature of their design. When the set clamping force is attained as the mould is closed, hydraulic pressure builds up in the clamping cylinder and is arrested at a certain predetermined level. This is consistent throughout the cycling of the machine so that clamping force applied is always the same. This optional extra therefore does not apply to machines of this type or to hydro mechanical models.

Mechanical scotch.

This is usually a bar which automatically drops down between the platens when these are fully opened to prevent them from closing should the electrical and hydraulic safety interlocks fail. Another type comprises a fixed bar and a stop plate or latch that drops down when the guard door is opened. A measure of protection is thus given to operators but the possibility of interlock failure is remote on a machine built to recognised safety standards. However, accidents do occur when operators work on machines with the guards removed and safety systems isolated and in such circumstances a scotch can provide a final resort.

Core pulling systems.

Mould requirements determine the need for one or more core pulling systems and provision for these in the machine's control system. A trade moulder may be advised to have at least two core pulling systems on each of his machines. As standard there may be none or only one. (See also page 161.)

Photo electric ejection.

A photo electric sensor on the ejection chute checks that the moulding has been properly ejected from the machine and inhibits the following cycle if not. This option is usually applied to moulding fairly large components which may be prone to fail to eject.

Air valve to assist demoulding.

Demoulding means extracting or taking out a moulding from the mould. A mould normally has a hydraulic ejection system with ejector pins or a stripper plate for this purpose, or may be demoulded by robot, but in the first place a moulding may have to be helped to be 'unstuck' or separated from the mould in order that it can be extracted. This is often the case with parts such as beakers, flower pots or buckets. An air valve is fitted to the machine so that a burst of air pressure can be applied at an appropriate point or points to separate it. This may also be sufficient to eject it as well, but in many instances the air blast and hydraulic ejection systems are both used, one being operated immediately after the other.

Where an optional air valve is fitted, the machine control system provides for the timing and duration of the blast and its coordination with the movement of the hydraulic ejector.

Mould loading device.

This is a simple crane comprising a pivoted jib and a block and tackle. It may be attached to the fixed platen and enables the mould to be lifted up from the floor, swung over the mould space and lowered into position for mounting.

Automatic tiebar retraction.

It may be necessary to remove an upper tiebar, or both upper tiebars, on a machine to mount a large mould or one with a number of protruding attachments such as core pulling cylinders. This is a lengthy job. Where the task is likely to be frequent, the machine can be provided with automatic tiebar retraction. A system of this kind involves hydraulically clamping the tiebar to the moving platen and then using the opening movement of the platen to pull the tiebar back. Forward platen movement is used to replace the tiebar. This feature is mainly offered for larger sizes of machine.

▲ General

Control system.

(see also page 95)

The standard control system will provide for the main functions of the machine but it may be necessary to specify the interfaces needed for certain items of additional equipment or facilities.
These may include the following: -

Cavity pressure transducer (see page 76)
Robot for part takeout
Gas injection (see page77)
Automatic or semi automatic mould change (see page 34)
Production logger (see below)
Statistical Process Control (see page 165)

Production logger software provides for the storage and retrieval of: -
Machine settings including stroke distances, speeds, pressures and temperatures for individual moulds
Part data (type of material, weights, shot volume)
Information for production and shift planning including machine and order status.
There are several proprietary production logging systems as well as those offered by individual machine manufacturers.

Hot runner control system.

A hot runner system ensures that the quantity of material in the mould channels between the entry point and the 'gates' to the individual cavities is maintained in a molten state, the object being to avoid producing scrap. The system is broken down into a number of zones each of which has an electric heater and a temperature control instrument.

A moulder may buy a complete package of proprietary equipment comprising the hot runner components that are fitted in the mould and a separate multi zone temperature controller contained in a box or floor-mounted cabinet. Alternatively it is possible to obtain the hot runner components only and use a controller optionally inbuilt as part of the machine control. Although this is a neater arrangement, a disadvantage may be that moulds without their own controllers can be run only on suitably equipped machines.

Temperature controller cut out.

This device cuts out the flow of cooling water through the mould if the machine is temporarily halted.

Anti vibration mountings.

A machine may be fixed to the factory floor by means of bolts grouted into the concrete, but anti vibration mountings can otherwise be used and are generally preferred. It may also be advisable to locate the machine with grouted pegs to prevent it from shifting from its proper position.

Star delta starter.

As an alternative to what is known as an 'on line' starter for the main motor, which may be standard, a star delta starter switches in the current in stages and avoids making a sudden high initial peak draw from the power supply.

Preheat timer.

This allows the reservoir oil and the barrel of the injection unit to be preheated in anticipation of the start of production, for example on a Monday morning.

Paint finish.

It is becoming more important that end customers be favourably impressed when they visit a moulding plant and having a number of machines in a variety of colour schemes detracts from this. Many users therefore have their machines finished in a house colour or to match the scheme of the make mainly represented in their factory. It is not uncommon to visit a machinery manufacturer and see a number of machines in competitors' colour schemes under construction.

Ancillary equipment.

A financial allowance additional to that for the machine and the mould should be made for the purchase of ancillary equipment. Many buyers do not do this and leave the purchase of ancillary equipment to the moment when the machine and mould finally arrive. As a consequence their choice is limited and they may be faced with the option either of acquiring the most expensive equipment or other of low price and questionable performance and reliability.

All machinery suppliers are able to quote for the supply of ancillary equipment. There are also a considerable number of companies who specialise in this field. Some are British manufacturers and others agents and distributors for equipment of foreign manufacture. Names and addresses will be found in the trade directories listed under 'Further reading' on page175.

The most important items of ancillary equipment to provide for a machine are a hopper loader and a mould temperature controller, although the hopper of a small machine may be replenished manually. Others that may be required include a material drier, a water chiller for mould cooling (in addition to the mould temperature controller), a robot and a belt conveyor for part take out, and a sprue picker and granulator for scrap reclaim. Although one may be fitted once a machine is delivered, a robot is a fairly specialised piece of equipment and is usually integrated with a new machine while it is being built, particularly a large type.

The factory in which the machine is installed will also require means for cooling water down to ambient temperature. This is normally a water tower and the oil reservoir cooler of the machine is connected to it. A water tower comprises a series of cascades and as the water pours over them its heat is dispersed into the atmosphere. This process may be assisted by a blower fan. It is possible to cool the mould off the same circuit but modern practice is to use a chiller with a higher rate of heat extraction. Cooling means reducing water temperature to ambient level: chilling is reducing it below ambient level almost down to freezing point and involves use of a positive heat extraction system.

A compressed air supply will also be required if the ejector system on the machine or the robot is pneumatically operated.

chapter 4

Further important points to consider. Screw performance. Proprietary components. Proportion of total shot volume which should be used. Shot weight repeatability and machine stability. Control systems. Closed loop injection control.

▲ Screw performance.

The most important factor in deciding which machine to buy is not mentioned in any table of machine specifications but must be assessed by practical trial. This is screw performance - how well the plastics granules which drop into the barrel from the hopper at one end are converted into an amount of melt of even temperature and viscosity at the other. Many purchasers of machines are unaware that their prime consideration should be the advantage that one machine may have over another in this respect. The ideal is termed a 'homogeneous melt', meaning simply that no solid or semi solid parts of granules remain and that the melt is of even viscosity throughout, with none of it thinner or thicker than the rest.

Viscosity and temperature are two sides of the same coin. A high temperature means a low (runny) viscosity, and a low temperature means a high (thick) viscosity. The purpose of the machine is to transfer the material, which starts off in the form of granules, into the mould, and the granules have to be converted into a flowable melt to make this possible. The process of plasticising is to put heat into the solid granules to turn them into a liquid melt, but whilst it is possible for any machine to do this some do it better than others and produce a melt of more even consistency. It is important for the buyer to make sure he is purchasing a machine which is wholly satisfactory in this respect because it will be far more profitable in use than one which is not.

A screw which is well designed and matched to a particular material will produce a homogeneous melt at the lowest possible temperature. A screw which is not so well designed will produce a melt some of which comprises low viscosity material and the rest high viscosity, and possibly even lumpy unmolten material called 'unmelt'.

It is impossible to make satisfactory mouldings if the melt is not homogeneous because the colder parts of the melt will solidify in the mould before the hotter. As they do and contract, they will pull other material, which is still in the molten state, from other parts of the mould cavity, stretching these in doing so. The entire moulding will eventually cool but the stretchings will be frozen in as unwanted internal tensions, or stresses. These will cause dimensional distortions and physical weaknesses in the moulded part.

When a homogeneous melt cannot be obtained at a certain temperature, the difficulty may crudely be overcome by raising the temperature of the process. The object of this is to thin out the thick, high viscosity parts of the melt and obtain a more uniform consistency. But the more heat that is put into the material, the more that has to be taken out during the cooling stage. This prolongs the cycle time and means that fewer parts per hour can be made. It should therefore be clear from this simplified explanation that good screw performance is fundamental to the profitability of the process and demands special attention by a machine buyer.

Some designs of screw are more effective at obtaining a homogeneous melt at lowest possible temperature than others and it is up to the buyer himself to find out which is best. No other feature of the machine is more important and even the most sophisticated machine control system cannot compensate more than partly for shortcomings in screw performance. Differences in screw performance do not necessarily coincide with how much one pays for a machine. The type of material that the machine is intended to process is the starting point of this line of investigation, not the price of the machine. An expensive machine does not necessarily have a better screw performance than a less expensive one unless the screw design is specific to the type of material.

Screw designs are in two categories: (i) general purpose and (ii) specific to certain types of material. It is chiefly in regard to the first category, general purpose screws, that differences between various makes of machine are likely to be found. Most moulders process a variety of materials and although the ideal is to use a screw specifically designed for each kind, this is impractical as the amount of downtime involved in changing screws between runs, if a change of material is involved, would be uneconomic. What happens in practice is that one material is purged out of the barrel and the next is run through it with least delay. A general purpose screw is used for all jobs and gives an acceptable level of performance for each even if it is not the ultimate best for any particular one.

The point here is that general purpose screws made by various machine manufacturers are not of a broadly similar universal design, as might be implied by the description, but differ from one make to another and cannot necessarily be compared. One maker's general purpose screw can perform quite differently from another's. *General purpose screws are not all the same.*

Individual manufacturers each offer a design of general purpose screw which they believe is the most appropriate for what they conceive to be their sector of the market. There are many moulders who process materials which are easily mouldable and make little demands on the machine because high standards of dimensional accuracy and surface finish, for example, are not required. This sector is chiefly served by makers of less expensive types of machine and these are perfectly good enough to make the products concerned. They do not need the features required for a fine degree of control over the process and the machine manufacturers serving this sector will supply general purpose screws designed for the materials which it mainly uses. These materials are known as 'commodity' plastics, that is they are used to make familar commodities.

Materials with higher requirements are processed in what might be termed the upper sector of the market. These are known as 'technical' or 'engineering' plastics and are physically stronger, have better appearance and are capable of being moulded to close dimensional tolerances. The demands of engineering plastics in regard to screw performance are not the same as for commodity materials and machine manufacturers supplying this sector provide general purpose screws matched to more stringent processing characteristics. Whilst these are still called 'general purpose' screws, their design is different from those intended for the more easily processed commodity materials.

Choice of screw largely turns on the types of material which one intends to process. A 'general purpose' screw for easily processed materials will give good homogenisation with these and be capable of high rates of output in terms of kg/hr. But it will not process more demanding materials as satisfactorily as a 'general purpose' screw designed for these. The converse is also true. A general purpose screw designed for engineering materials will not be as effective with easily processable grades as one specifically intended for these and will have a lower kg/hr output.

Going a little into the technicalities involved, the difference between the two types of screw is chiefly in respect of what is termed the compression ratio. This is the ratio between the height of the screw flight at the feed section, where the material drops in from the hopper, and its height in the final metering section. A high compression ratio screw may be used for easily processable commodity materials for which a high rate of output is required, and a low compression ratio screw for sensitive technical materials that require careful processing.

A general purpose screw offered by a manufacturer serving the sector processing less demanding materials may have a compression ratio of about 2.4:1, but a manufacturer supplying the sector running more critical materials will opt for a general purpose screw with a compression ratio of perhaps 2.0:1. Thus there are two designs of screw both of which are described as 'general purpose' but with a considerable difference between their compression ratios. It follows that their performance will differ. One is not necessarily better than the other, it is simply that each type has to be carefully chosen in relation to the materials which are to be processed. Practical trials are the only way to find out which manufacturer offers the most satisfactory type.

Compression ratio is also expressed in terms of material volume in the respective sections of the screw and in any discussion on this point with a supplier it is as

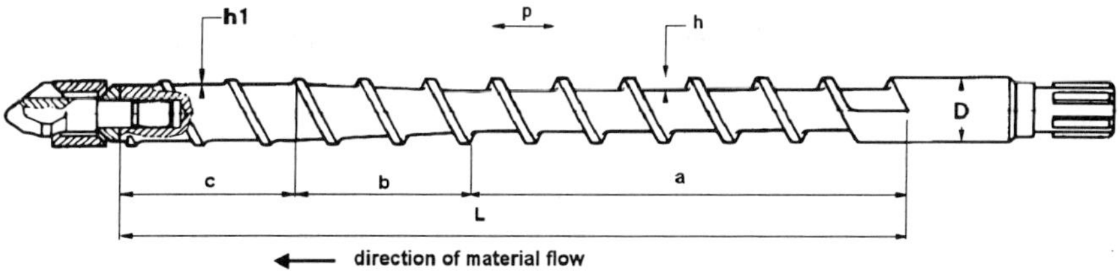

Drawing of a typical screw of an injection moulding machine, its essential part. The screw rotates within the barrel to melt the granular material and then makes a forward movement under high pressure to inject the material into the mould. It has a helical flight or thread for most of its length and this pushes the material along by an Archimedean action. The height of the turns of the flight gradually diminishes so that the material is compressed into a molten condition.
L = Length (of effective flighted section).
D = Diameter.
p = Pitch of flight (length of one turn, usually equal to diameter).
a = feed section where material enters from hopper and is initially warmed up.
b = compression section in which the height of the turns of the flight reduces.
c = metering section from which the molten material is transferred through the shut off valve to the front of the barrel.
h = initial flight height.
h1 = final flight height. The ratio of h to h1 is the compression ratio and is made to suit the material or class of materials for which the screw is intended.
Screw performance is the most important factor to determine in choosing a machine. Performance is largely affected by the Length/Diameter (L/D) ratio and the compression ratio. The L/D ratio is that between the length of the flighted section (L) and the diameter (D). Most machines have screws with an L/D ratio of 20:1, but 25:1 and 28:1 are also offered. The higher the L/D ratio, the greater the amount of plasticising treatment that the material receives. The more widely used materials withstand a high compression ratio but technical or 'engineering' materials do not. This means that a screw designed for one kind of material will not give optimum results with another. The ideal is to obtain an even melt viscosity at the lowest attainable temperature and in the shortest possible time.

well to find out whether one is talking in terms of flight height or volume.

A special purpose screw, as distinct from a general purpose, is strongly to be recommended where only one or a limited range of materials with almost identical characteristics is to be processed. For example, a number of moulders serving the electrical appliances market run nothing else but ABS on many of their machines and there are special purpose ABS screws from nearly all machine manufacturers which provide the optimum performance for this material, particularly to give high surface gloss. Even here, though, some manufacturers' ABS screws are held to be better than others as a result of the amount of design and development that has been put into them, and the buyer should seek the make of machine which gives the best performance.

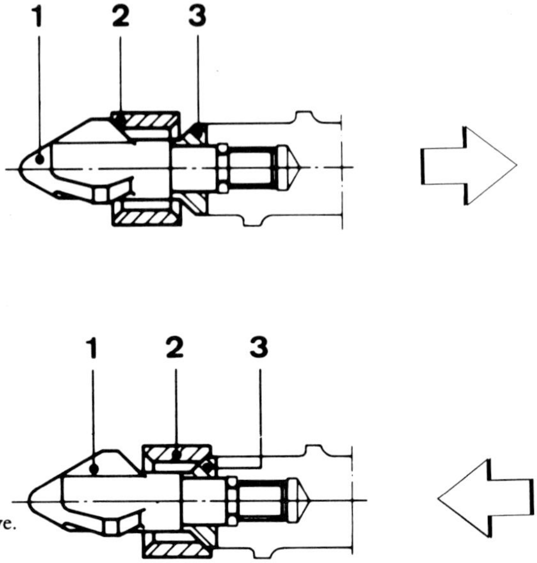

Screw shut off valve.
1 Screw tip.
2 Check ring.
3 Check ring seat.
The shut off valve at the front end of the screw is usually of the check ring type as shown here. The upper diagram shows the check ring in the position it takes during plasticising. The screw rotates, moves back and plasticised material flows through the check ring and slots in the tip to the front of the barrel.
On injection, the screw moves forward to force the melt at the front of the barrel out through the nozzle, but it is necessary to close the passage through the check ring as material would otherwise leak back. The check ring therefore automatically moves back against the seat as the screw goes forward and cuts off the flow (lower diagram). Accurate shut off valve performance is necessary to produce mouldings of consistent weight and this may be assessed during a machine trial.
The complete valve can be unscrewed from the end of the screw for cleaning. If dirt such as particles of scorched material or foreign bodies which find their way into the melt should stick to the valve seat, it will not completely shut and moulding cannot be properly performed.

Manufacturers compete strongly against each other to produce the most efficient screws for particular materials and devote a great deal of their R&D to this. Progress tends to be non-uniform over a number of makers. At any one time, for example, maker A will enjoy an outstanding reputation for the performance of his special purpose screw for ABS or other material, but the position can change quickly when rival maker B brings out a new and improved design and overtakes A. For the machinery buyer it is matter of combing the market at any particular time to find the best available type of screw.
The design and performance of the screw shut off valve is also of great importance in achieving consistent moulding quality. This is at the front end of the screw and is usually in the form of a ring or collar that slides forwards and backwards over a short distance. There are other types that have tubular passages for the flow of material and a ball in each passage moves to permit flow or shut it off.
Screw shut off valves are not to be confused with nozzle shut off valves (see page 75).

▲ Proportion of total shot volume which should be used.

In a preceding chapter, it was stated that the shot volume of a moulding should not exceed 60 or 80 per cent of the maximum available and that this factor had to be taken into account in choosing an injection unit of suitable size. The reason for this will next be explained.

It is good moulding practice to minimise the time that material stays in the barrel once it has been plasticised until when it is injected into the mould. This is called the 'dwell' or 'residence' time. During this period the material absorbs heat from the barrel and if dwell time is prolonged the material will overheat and reduce in viscosity by too much. If an exceptionally large shot is plasticised, the material which enters the front part of the barrel ahead of the screw stays there for considerably longer than the last amount to arrive. If, for example, the plasticising, or screw rotation, time is 20 seconds, the first amount of material will dwell at the front of the barrel for up to 20 seconds longer than the last. During that time it will absorb more heat from the barrel and its viscosity will be lower. This difference may be critical. (This is a simplified explanation. Other technical factors are involved.)

It is undesirable to inject a shot with an excessive difference in melt temperature, and hence viscosity, from one part to another because a moulding with internal stresses liable to cause it to distort will be produced. The difference in temperature from one part of the shot to another is called the 'melt temperature profile' and the maximum permissible may be only 1 or 2 degrees C, according to the type of material. Some materials are less temperature tolerant than others. The highest proportion of available shot volume to use for high quality moulding should be 60 per cent and for less critical production 80 per cent. There is, for the same reason, a limit to the minimum amount of available shot volume that should be used, 40 per cent being recommended for top quality moulding and 20 per cent for other work. (Special conditions may be involved in technical moulding and this advice will not always apply.)The total amount of material in a barrel around the flights of the screw and at the front is called the 'inventory' and it can be appreciated that if the shot volume used for a moulding is small it is only a low percentage of the inventory. In turn this means that the material making up the inventory will stay in the barrel for a long time before it is finally injected into the mould. Total barrel inventory is about twice the maximum shot volume.

Heat is absorbed during the residence time and as a result the material may degrade, or oxidise, affecting the quality and strength of the moulded part. It is also undesirable for the shot volume to be too small because the power consumption of the machine will rise steeply. When related to the amount of material processed, this factor is known as the 'specific power consumption' and an optimum figure is 0.8 kW/hr per kg of material. Specific power consumption can rise to 2.5 kW/hr per kg or even more if the percentage of shot volume used is too low.

Equally one may improve specific power consumption considerably by using more than the permissible maximum percentage of available shot volume, but the likely penalty incurred in producing poor quality mouldings would make this false economy.

▲ Shot weight repeatability and machine stability.

As well as seeking a machine with good screw performance and optimum shot volume capacity, a buyer should ensure that it is capable of maintaining good shot weight repeatability. A machine and mould are initially set up to produce mouldings of a predetermined shot volume and therefore weight, and good shot weight repeatability ensures that this volume is maintained within certain limits, or 'tolerances'. Also, the better the shot weight repeatability of the machine, the closer together that the tolerances may be set and material saved.

A further factor that arises here is called machine stability. A high standard of stability is a prerequisite of good shot weight repeatability and a machine is said to have good stability if it maintains its cycle time and the quality of its output within close limits over a period of operation of 24 hours or more.

Cycle time variations are usually related to changes as between day and night that occur in ambient temperature and the temperature and pressure of water used to cool the mould and the oil in the machine reservoir. A machine with good control should be less affected by such changes and be more stable than one with a less refined control system, although the design of the hydraulic circuit of the machine is also relevant. The hydraulic circuit of a poorly designed machine may, for example, be more prone to overheating than a well designed machine and this will affect how it maintains its cycle times and shot weight repeatability.

These diagrams illustrate why good shot weight repeatability is important. The upper diagram represents a number of mouldings made on a machine with poor shot weight repeatability. Three mouldings are outside the limits and may need to be rejected. The lower diagram represents the performance of a machine with better shot weight repeatability. All parts are within the weight limits. A good machine in this respect allows mouldings to be made at a lower average weight and for material to be saved.

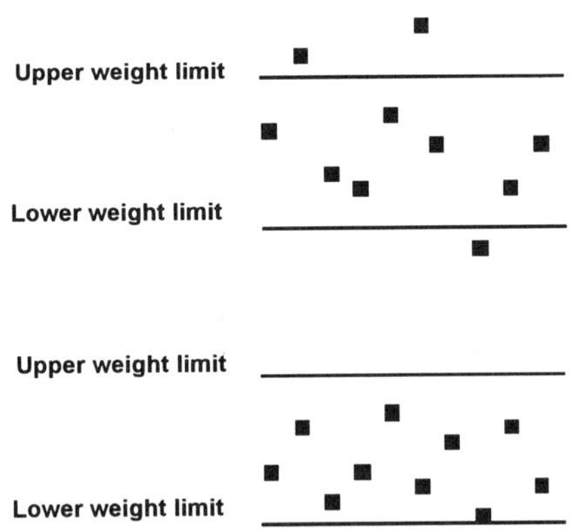

SHOT WEIGHT REPEATABILITY

Overheating causes the temperature of the hydraulic oil to rise, which will reduce its viscosity, and this affects the speed of movements made by the different parts of the machine during the cycle.

These considerations can be tested only by a practical machine trial (see page138).

▲ Control systems.

Control systems of modern machines provide for centralised control to a greater or lesser extent. Before electronic controls were introduced, a number of manual settings for stroke distances, speeds, pressures and temperatures needed to be made to valves, switches, regulators and other control devices on different parts of the machine. Initial setting and subsequent resetting therefore involved going around the machine to make the various adjustments.

Electronic control enabled setting to be centralised so that all, or nearly all, adjustments are made from one point, that is at the control panel. Moreover settings can be memorised on a disc, cassette or other type of data storage carrier. The expense of a centralised control system is, however, proportionately high for a small machine and simplified versions for these are offered at lower cost. They provide for centralised setting of certain values at the control panel but switches and other regulators are fitted on the machine for various other functions and need to be adjusted manually. This compromise arrangement is usually acceptable because of the lower price that is asked for the machine.

There are generally two or three levels of control system offered for a machine of a certain size. The bottom level, at the lowest cost, may not provide full centralised control but require that some settings be made manually. The next level will be more comprehensive and offer greater refinement. For example, injection pressure may be adjustable in three steps with the bottom level control but in five with the more expensive. At the top level, the system will probably include facilities for closed loop injection control (see below), production monitoring, SPC (Statistical Process Control) and connection to a mainframe computer.

It is worth while to check whether fault location is included in the facilities of the standard control system of a machine under consideration as this can save valuable time. A typical feature for this depicts a diagram of the machine. Lights are illuminated to show the sequence of the various valve and other operations performed during the cycle. If a fault occurs and the machine stops, the location of the fault is apparent from where the sequence of the lights has halted. Some makers offer a modem link so that fault location can be done remotely, allowing the distributor's or even the factory engineers to correct a user's machine.

The power of the microprocessor incorporated tends to increase the higher the level of the control system and the latest are of the RISC (Reduced Instruction Set Computer) type providing extremely high response speeds and almost limitless memory. These are up to 20 times more powerful than older systems, which have come to be known as CISC (Complex Instruction Set Computer).

Where there is a choice between colour and monochrome VDU (Visual Display Unit) screens, the colour option is more expensive but is advantageous for graphical display. There are also EL (Electro Luminescent) screens which are monochrome and, being flat, tend to pick up fewer reflections. These provide a compact and finely detailed display.

A few machines have touch sensitive screens which are said to be easy and quick to use, but input of data on most types is by means of a membrane alphanumeric keyboard. The keyboard may be supplemented with a trackball. On a machine with such a feature, the injection stages are not set by entering numbers from the keyboard but by drawing a graph or profile on the screen with the trackball.

Symbols rather than words are employed with some systems to represent various machine functions. This may be of benefit where there are personnel not fluent in English, although these systems are also claimed to be easier than those of the keyboard type for literate persons to use.

Broadly, manufacturers have sought to simplify the MMI (Man Machine Interface) at the same time as refining the control systems of their machines and extending the range of functions they can perform. These are conflicting aims and as a result the trend is towards 'black box' or 'push button' technology, which largely means taking the control function out of the user's hands, because it becomes too complex to be easily understood, and putting it into a system programmed to perform in a certain manner according to input values. These 'expert' or 'intelligent' systems relieve the user of having to judge for himself the settings that are needed, but provision is made for fine tuning so that the automatically implemented programme can be precisely adjusted to specific needs.

Manufacturers tend to give specific brand names to their control systems in addition to the names or designations they give to the machines themselves. This practice arose following the introduction of electronic control systems, which replaced the use of electro mechanical devices. It became possible to integrate the sequence of the phases in the machine cycle and for these to be run as a predetermined programme rather than as a succession of individual events, the end of one phase initiating the start of the next. Old machines had controls but not control systems in the modern sense.

As controls became systematised, they acquired identities in themselves and the next step in their evolution was for them to be given names. This process has, however, become a little out of hand and tends towards confusion. Not only do manufacturers give names to their control systems but many give a different name to each of their systems at various levels. There are more names for control systems than there are makes of machine. Buyers should ensure that they understand exactly what features are offered with a particular control system, irrespective of its name, in comparison with others and define whether the interfaces and outputs that they need are provided either as standard or optional extras. This may entail going through the full specification of a control system item by item, even although this may be more detailed than the specification for the machine itself.

▲ Closed loop injection control.

Offered by a number of manufacturers either as standard or as an option for their top of the range models, so-termed closed loop injection control is a means for ensuring that the melt is injected into the mould at as consistent a speed as possible so that best quality moulding can be attained. It can be likened to cruise control on a car which enables a set speed to be continuously maintained irrespective of road gradients.

Closed loop injection control is provided on most machines with this feature by what is known as a 'servo' valve which regulates the actual injection valve. 'Servo' (literally a slave) means to serve or assist: the action of the servo valve assists that of the injection valve. Most machines have hydraulic circuits of what is called the proportional type but some have digital. It is beyond the scope of this book to explain the difference other than to state that closed loop injection control applies only to machines with proportional hydraulics. Makers of machines with digital hydraulics achieve an evenness of injection speed by other means.

The term 'closed loop' means that there is a feedback system whereby the speed of the part of the machine concerned, in this case the spool of the injection valve, is monitored, that is its actual position is reported back to the microprocessor every few milliseconds. The spool is the internal part of the valve which controls the flow of oil through it, just as a bathroom tap may be opened more or less to control the amount of water which flows through. The signals which report the position of the spool are compared within the microprocessor against a predetermined program and its position is adjusted as necessary to keep it as near as possible to where it ought to be. This adjustment is done by the servo valve.

As mentioned on page 62 under 'Maximum injection speed', where an extremely high injection speed is set, possibly to be performed in less than 1 second, the screw will not in fact reach the set speed as there is insufficient time for it to accelerate up to it. This being so, it will not come under closed loop control even if this is activated. Microprocessor control systems allow set and actual speeds to be seen on the screen and these can be observed. Over extremely short injection times, actual injection speeds may not even reach 60 per cent of the rate set. This means, for example, that if an injection speed of 800 mm/sec is set and is to be controlled in closed loop but the screw never in fact goes faster than 500 mm/sec, the closed loop system is not called upon to perform its task. At the end of the fast filling phase, the holding and recovery phases may, however, be valuably controlled in closed loop, as discussed below.

In the absence of closed loop control, the spool can be affected by inconsistencies in the viscosity of the melt. Melt may be inconsistent for a number of reasons. The rate at which granules fall into the barrel may be erratic or there may be moisture in the material. There may also be variations in material quality as originally supplied. On injection, the screw moves forward to force the melt through into the mould, but it will be appreciated that if the melt varies in viscosity the screw cannot maintain an even speed. At one moment it may be pushing against a thick, high viscosity amount of melt and thin, low viscosity the next.

This causes it to move forward erratically, which makes matters worse, and it may even shuttle backwards and forwards. In turn the spool of the injection valve reacts to the jerky movement of the screw and wobbles about. The object of closed loop control is to diminish the extent to which the spool will behave in an irregular manner and this in turn moderates the behaviour of the screw. In effect the movement of the screw is stabilised, not completely but to a considerable extent, and the production of mouldings which demand high melt quality - Compact Discs for example - is made possible.

It might be asked why the screw itself is not controlled in closed loop rather than the injection valve spool. The answer is that machine microprocessor control systems are not yet powerful enough or sufficiently quickly responsive for this to be done. In the meantime regulating the spool is a compromise but an effective one.

Where closed loop injection control is provided, holding pressure and screw back pressure can also be controlled in closed loop as the same valve governs all three functions. Holding pressure is the pressure applied by the screw against the material injected into the mould to prevent it from squirting back and out again. Molten material is elastic and will 'spring' out of the mould cavity unless prevented from doing so.

The concept of viscoelasticity, the elasticity of a viscous substance, needs to be grasped in order to understand how a plastics melt behaves. This characteristic may seem odd because most of the liquids with which we are familiar in the course of our daily lives do not exhibit viscoelasticity. Nevertheless plastics melt is elastic, rather as putty or dough, and can be compressed by about 10 to 30 per cent.

This compressibility helps the material to fill the mould cavity. The melt is compressed and precharged with an amount of stored energy before it is forced into the cavity where, upon arrival, the precharged energy is released like a spring and the quantity of melt expands to fill the cavity void. This release of compression augments the injection pressure exerted by the screw, but the material would spring back and expand again if the force tending to compress it, that is the pressure exerted by the screw, were to be completely removed. This is why holding pressure must be maintained against the melt by the screw after it has been forced into the mould. Holding pressure (sometimes also known as 'packing pressure') is taken off once the moulding has cooled and solidified.

Holding pressures for commodity materials are usually between 250 and 600 bar and for technical materials 400 and 800 bar. Although the first stage of injection pressure is normally at a higher level than the second holding pressure, the reverse may apply to thick section mouldings. Only a low first pressure may be needed for a moulding more than 3 mm thick, for example, but the holding pressure required to prevent the melt from escaping could be higher than in the first stage. In such cases, it is necessary to check that the machine control permits this.

With materials known as 'amorphous' and which solidify gradually, it is also possible to diminish holding pressure progressively as the moulding cools instead of all at once, and this is indeed advanced injection moulding practice observed by top class moulders. It is not necessary to have a high holding pressure at the end of the cooling period when the moulding has almost completely solidified because the tendency to push the screw back gradually diminishes as the moulding cools. In fact it is probably undesirable to maintain a high holding pressure for the whole of the cooling time because it becomes greater than required and may cause the moulding to be distorted.

This principle underlies what is known as 'PVT' moulding, the letters standing for Pressure, Volume and Temperature, three interrelated variables. A buyer may check that the control system of a machine permits holding pressure to be easily regulated according to this principle. Mouldings in what are termed 'crystalline' materials change from the molten to the solid state quickly and full holding pressure is maintained until they do so.

It is generally agreed by authorities on the subject that the dimensional accuracy of a moulding, and consistency in this respect from cycle to cycle, is considerably influenced by the holding pressure, and closed loop control enables this phase of the cycle to be well regulated. It is, as mentioned, also of value during the recovery phase. During screw rotation, or recovery, the screw works itself backwards as material is transferred from the hopper end of the barrel (known as the back) to the front. As the volume of material at the front increases, the screw is displaced and pushed backwards. However, if there is no pressure applied at the back of the screw it will be pushed backwards against no resistance. It is desirable with some materials for there to be some resistance and where this is applied it is called 'screw back pressure'.

In the case of a highly viscous material such as polycarbonate (PC), for example, a certain amount of back pressure is needed as the tendency for the screw to be pushed backwards is greater than with a low viscosity material. The application of screw back pressure gives the screw something to push against as it rotates and plasticises the material. In the absence of back pressure the screw may slip, meaning that it will travel backwards quickly and freely over a certain distance instead of at a steady, controlled rate. This is undesirable.

Slip may also occur rotationally, meaning that the speed of rotation of the screw will increase for a brief burst. This is also undesirable but can be avoided by the application of back pressure. Closed loop control of back pressure enhances the performance of the screw during the recovery phase and assists in obtaining good melt consistency, although achieving a consistent rate of material feed from the hopper and good temperature control in the area of the hopper throat are also important.

Three phases of the moulding cycle, filling the mould, applying holding pressure and screw rotation for recovery, may thus be controlled in closed loop. Although the advantages of closed loop injection control are mainly emphasised in relation to the first of these, its value in respect of the other two is nevertheless important. Some authorities in fact believe that closed loop control of holding pressure is more important than during mould filling.

Closed loop injection control is standard on some high grade makes of machine and an option on others. Where it is offered as an option, it also entails specifying the highest control system available as the lower systems will not have sufficient capacity for the regulation of the servo valve. This option is therefore relatively expensive for the smaller sizes of machine. Good machine stability, as discussed above, is a necessary basis for closed loop injection control.

▲ Proprietary components.

Machinery manufacturers tend to use hydraulic and electrical components from a limited range of manufacturers. Although machinery distributors stock spares of proprietary hydraulic and electrical components, the prices they charge are high and replacements can often be more cheaply obtained from official equipment stockists. Experienced machinery buyers prefer machines which incorporate proprietary components of well known and widely stocked makes rather than lesser known makes which may be difficult to obtain.

Nevertheless, a stockist may not always be able to give immediate delivery of a particular item because he is supplied by the manufacturer concerned on a batch basis. When a batch is sold out, he cannot supply any more of the components concerned until the next arrives, which will depend upon the manufacturer's production schedule and may be indefinite.

To enable users to avoid long production hold ups on this account, a machinery distributor undertakes to hold a guaranteed spares stock, guaranteed meaning that he can ensure being able to supply any particular component within 24 hours. Because this ties up a considerable amount of his capital, the distributor must price his spares stock accordingly, which is why the cost of buying spares from a distributor is high. However it is not when weighed against the idle time of the machine and production lost. A distributor for a leading make of machine may have at least £1 million of spare parts in stock for immediate delivery.

chapter 5

Different injection moulding machine designs. Reciprocating screw and other types of injection unit. Toggle, direct hydraulic, toggle/direct hydraulic and hydro mechanical machines. Tiebarless machines. All electric machines. Multi colour and multi component machines. Multi station machines. Double daylight machines. Machines for unusually large mouldings. Rebuilt machines. Low shot weight machines. Manual and semi automatic machines.

This chapter comments upon the major points of injection moulding machine design to assist the prospective buyer, and it is not intended to be an exhaustive review as may be found in a comprehensive technical work on the subject. The reader is also referred to the Glossary, page 159-170, which contains descriptions of machines for special applications.

▲ Injection units.

Almost all modern injection moulding machines are of the reciprocating screw type. A brief historical review makes this term clear. Early machines of the 1930s had a simple cylinder into which the material was fed and electric heaters supplied heat to melt the material. Injection into the mould was by means of a plunger, or piston. These were therefore known as plunger injection machines.

They consumed electricity heavily and the invention of the screw greatly improved economy. A rotating screw in effect vigorously stirs the material, creating heat by friction and shear ('kneading'), and this causes the material to melt. The amount of current needed to drive the screw round to achieve this is

much less than simply by using electric heaters.
The first screw machines, available 20 years later, had a screw to melt, or plasticise, the material and this was transferred into a cylinder for injection by a plunger. Actual injection was therefore performed as it had been on the early machines. The first stage of melting by means of a screw was known as preplasticisation.
The later invention of the reciprocating screw simplified the design of the injection unit because the screw combined the function of plasticising the material with that of injecting it.
As the screw rotates, it conveys material to the front of the barrel by the Archimedean, or 'propellor', action of its flights. Material accumulates at the front and as it does so it pushes the screw back. (The injection unit is designed to allow this.) The extent to which the screw is pushed back is adjustable according to the shot volume required. The screw stops rotating once the shot volume has been plasticised and it is next pushed forward under hydraulic pressure to express the melt out through the nozzle. This is the injection stroke, and once it is completed the screw is back in the same position as it was before it began to rotate. It thus reciprocates, or moves backwards and forwards.
Heaters on the outside of the barrel keep it up to temperature. There may be an interval between the time when the screw stops rotating and when it moves forward again to make its injection stroke (or between when it has injected the melt and starts to rotate). Because it is not rotating, it is not putting heat into the material and the function of the heaters is to ensure that the barrel does not cool down too much.
The reciprocating screw injection unit has not entirely supplanted the older designs. Small pneumatic benchtop sized machines with plunger injection units are manufactured. These are low priced machines which would be too costly to make with a screw injection unit. There are also machines with a screw preplasticiser and a plunger injection cylinder to make blow moulding bottle preforms in PET (polyester), for which purpose this design has certain advantages. Some small machines for high precision moulding are also of this type.

▲ Clamping units.

Toggle machines.

A toggle is a series of links with hinged joints. It can be fully extended in a straight line or collapsed and folded zigzag so that its ends are brought closer together. A human arm hinged at the shoulder, elbow and wrist is a toggle. An oarsman stretches out his arms and then folds them back to pull the oar up to his chest. This action is the same as that of a double toggle system of an injection moulding machine to make the moving platen travel backwards and forwards.
Toggle machines form the largest group on the market and in use. Most are double toggle designs but there are some small single toggle (one armed) models.

Points to look for on a toggle machine are robust links and generously sized joints. The joints contain short pins, corresponding to the pins in a simple door hinge, which take the strain when the toggle is extended and clamping force applied. The larger the diameter of the pins, the better they will take the strain.

Lubrication of the joints is crucial to the satisfactory operation of a toggle machine. At the simplest level there may be grease nipples for this and a grease gun is applied at stated intervals. Automatic lubrication systems are strongly to be recommended. Most machines have a centralised system with a reservoir of lubricant - either oil or grease - and small bore copper or nylon tubes leading from it to the individual joints. Such a system may either be fully automatic or manually operated at specified intervals. If copper pipes are fitted, they should be inspected frequently to check that they have not accidentally been crushed. This could block the flow of lubricant and cause a breakdown or seizure of the joint. Seizure can occur within an hour of lubrication failure.

The joints are the parts of a toggle system with greatest liability to wear and failure, which is why some users prefer the direct hydraulic design. By the nature of their design toggle machines can be made to run much faster than direct hydraulic machines, and a toggle machine designed and built for speed will be more expensive than a direct hydraulic.

A toggle system may have either four or five joints, or 'points', linking its separate sections. Four point machines have now largely given way to five point types. The advantage of the five point design is that it can be made to fold back within a shorter distance and this provides a longer platen stroke within a given overall length. Some energy saving is also gained. Some of the four point machines available are of Far Eastern manufacture and based on older Japanese designs made obsolete by their makers in favour of five point types. There is no reason why, within their limits, they should not be good value for money but their shorter stroke compared with five point may be restrictive and allow only relatively shallow mouldings to be produced on them.

The characteristics of a toggle clamping unit may be considered in relation to the shape and size of the article to be moulded. When extended, the arms of a double toggle system act against the four corners of the moving platen and this tends to spread the clamping force equally over the area of the platen. A large area moulding such as a tray, or a multi cavity mould in which there are several rows of individual cavities, should benefit from the good distribution of clamping force of a toggle machine. A toggle machine also inherently provides greater mould opening force.

On the other hand, because clamping force is applied towards the corners of the platen, the centre is less well supported than with a direct hydraulic machine of the type with a ram acting against the centre of the moving platen. Each toggle connection to the platen applies a quarter of the total clamping force and, because of the deflection of the platen (bending, in simple terms) when material is injected at high pressure, the platen will tend to bow in the middle. This deprives the mould of support in the centre and flashing may occur, dependent upon the configuration of the moulding in this area. A direct hydraulic machine may therefore be preferred for mouldings which require good central support, an example being a deep container or vessel such as a pail.

Irrespective of the type of clamping system, a machine should have substantially proportioned platens and different models should be compared in this respect. The tendency for a platen to deflect under load either in the centre or nearer the corners, according to the type of clamping system, is largely offset by well designed and substantially proportioned platens and, in practice, large area mouldings are satisfactorily made on direct hydraulic machines, also deep containers on toggle machines, where the platens are of adequate strength.

To a great extent the question of whether mould support is adequate or not is unrealistic so far as well designed machines of any type are concerned and will in any case be resolved by a practical machine trial. Machines of any type with all too obviously flimsy looking platens should be avoided even although they may be low in price.

Direct hydraulic machines.

These are so-called because the hydraulic cylinder or cylinders for the clamping unit act directly against the moving platen instead of via a toggle system. A toggle machine has a small clamping cylinder to operate the toggles and the toggles in turn move the platen. The clamping cylinder of a direct hydraulic machine is considerably larger and is connected directly to the moving platen.

There are two main types of direct hydraulic machine. The more common type has the clamping cylinder behind the platen and the ram acts against the centre of the platen. The front part of the ram has a cut out section in which the hydraulic ejection cylinder is housed. A variation on this layout is where there are two clamping cylinders side by side with the ejection cylinder positioned between them. Clamping cylinder diameters tend to differ between makes of machine, some models tending to have extremely large diameter cylinders in relation to platen size. These might be expected to give a good spread of platen support.

The other main type of direct hydraulic machine has four cylinders mounted on the fixed platen on the same side as the injection unit, an arrangement which reduces the overall length of the machine and saves floor space. An advantage of some models of this design is that the mould space overhangs the end of the frame of the machine and there is better access to it for handling the mouldings once they have been ejected. Clean room operation may be easier as the clamping units can be arranged within the air conditioned clean room and sealed off, with the rest of the machine outside in a normal environment. The warmest part of the machine is therefore outside and air conditioning costs are reduced.

With this design, the clamping cylinders pull the moving platen at each corner in order to apply clamping force. This is therefore applied not at the centre of the moving platen, as with the other design, but at the corners and in this respect there is a similarity with a toggle machine.

The simplicity of the design of direct hydraulic machines and the ease with which they can be set up appeals to many users. A toggle lubrication system is unnecessary and there is no susceptibility to joint and linkage wear. There is no separate mould height adjustment with most machines of this type as the available stroke length provides for this.

The chief attraction of direct hydraulic machines is that clamping force is maintained consistently cycle to cycle. Hydraulic pressure is sensed in the clamping cylinder at a predetermined level and this corresponds to the clamping force applied. Excessive clamping force cannot be applied because the pressure in the clamping cylinder cannot exceed a certain level. On a toggle machine, clamping force is set according to the distance between the platens to which the machine is adjusted for mould height. If the mould should expand with a rise in temperature, the clamping force acting upon it increases because the mould height adjustment is unalterable (except where automatic adjustment is provided - see page 83).

Toggle/direct hydraulic machines.

Some makes of machine combine toggle and direct hydraulic systems. The toggles are employed to close the mould and the direct hydraulic system, acting from behind the extended toggles, pushes these forward and applies the clamping force.

Hydro mechanical machines.

In the heavy machine class, some moulders prefer toggle machines, particularly for large area mouldings which require a wide spread of clamping force behind the moving platen. Mouldings of this kind are usually fairly flat and do not call for a long opening stroke. Where this is not the case and the mouldings, such as deep pails, boxes or other containers, do need a long opening stroke, the machine designer has to meet the requirement without building an enormously long machine.

The solution is to design a machine with a two stage clamping system and these are known as 'hydro mechanical' or 'lock and block' types. There are many variations but essentially they all work in a similar manner. The clamping unit is basically of the direct hydraulic type but in effect comprises two systems, one of which telescopes within the other instead of there being a single system of considerable length.

In the first stage of mould closing, one system extends the platen forward for a certain distance. The first stage system is then locked into its extended position by some positive mechanical means. On most machines there is a chock, or 'abutment', which is placed behind the ram or rams and on others a mechanism to secure the moving platen to the tiebars. The second stage system then provides the final amount of platen movement and applies the clamping force. When the mould is opened, the second system retracts first. The locking abutment or mechanism is then released and this allows the first system to be retracted.

Hydro mechanical machines usually have a means for mould height adjustment as found on toggle machines. This system is set so that the first stage system can be fully extended and only a short stroke is made by the second stage.
Some machines of this type have the two stage clamping system entirely behind the moving platen, whilst others are variants of the design with four clamping cylinders on the injection side of the fixed platen. Other machines completely disconnect the tiebars from either the fixed or moving platen to allow the moving platen to retract freely for a long distance. This enables exceptionally long mouldings, which need an unusually lengthy mould opening stroke for ejection, to be produced. With machines of this type the platen does not travel on the tiebars but on guide rails or slides in the base of the machine.
Hydro mechanical machines save floor space but, being of fairly complex design, are relatively expensive. At one time they were made in only the very largest sizes but improvements in design have enabled them to be offered in the medium range from 4,000 kN. A disadvantage is that cycle time is increased because of the need to actuate the mechanism which locks the first stage on mould close, and unlock it again on mould open. This was for many years a major problem but improvements were made and hardly any single stage direct hydraulic machines remain above this size.
The main drawback of big direct hydraulic machines was the necessity to pressurise all the oil in the clamping cylinder to apply clamping force. With a hydro mechanical design, only the oil contained in the second stage cylinder needs to be pressurised as the first stage is mechanically locked up. A considerable volume of oil needed for the operation of the clamping unit is saved, in turn enabling the size of the hydraulic reservoir and its oil cooling requirements to be reduced.
Some large hydro mechanical machines are built as two machines placed side by side. Their fixed platens are lined up to present a single surface for the fixed mould half and likewise their moving platens are coupled together to act as one. A 60,000 kN machine, for example, will comprise two 30,000 kN machines connected together so that big objects such as boat hulls can be moulded. When not required for such work, each constituent machine can be operated independently.

Tiebarless machines.

Tiebarless machines with vertical clamping units have been made for many years and are known as C frame machines, an obvious analogy. The horizontal tiebarless machine is a C frame machine turned through 90 degrees so that the clamping unit operates horizontally instead of up and down, but the frame is considerably reinforced to resist distortion. This arises when clamping force is applied. Absence of tiebars has advantages in improving access to the platens for mould changing, installing automation devices such as takeout robots, and allowing maximum use to be made of platen area.

With a conventional tiebar machine, the tiebars themselves are obstructions to full use of platen area. Where a large platen area is required, a tiebarless machine may therefore provide this at a lower clamping force, and price, than a tiebar machine, but a comparison with the platen dimensions of a tiebar type machine should be made to check that this is in fact the case.

The clamping systems of tiebarless machines available are of the direct hydraulic type and the design has been developed up to models of 4,000 kN. Machines without upper tiebars have also been introduced. The lower pair is retained but absence of the top pair improves access to the platen area.

▲ All electric machines.

With the development of modern servo electric motors, machines which are entirely electrically driven and dispense with a pump, valves, cylinders and other hydraulic components are being introduced by various makers. They are not new in principle and in fact in the 1930s electric machines preceded hydraulic, but the types of motor available at that time were simple compared with modern designs. When it was introduced, the hydraulic machine was demonstrably superior and has since maintained its dominant position.

The main advantages of the new types of all electric machine are exactness of control and low energy consumption. Chiefly because individual motors are needed for the various machine functions, all electric models are more expensive to build than hydraulic and generally cannot be justified exclusively on the grounds of their energy economy. The difference in price between the two, however, is likely to diminish with further electric motor development. A buyer may either investigate the applicability of an all electric machine for its specific attractions irrespective of price or wait for the prices of these types to fall.

Absence of tiebars on a tiebarless machine allows the whole of the platen area to be used for mounting a mould and simplifies the use of a robot for part take out. This picture shows the platens of a 4,000 kN machine of this type.

The standards set by the first modern all electric machines stimulated the improvement of the energy economy and other aspects of the design of conventional hydraulic models and these will maintain their cost/performance advantage until the prices of all electric machines become more attractive. All electric machines cost about the same as equivalent top of the range hydraulic models.

Peak current draw of an electric machine is lower than for a hydraulic machine and this may be considered where penal tariffs apply should a kVA limit be exceeded. A further advantage is that no oil is used and no provision has to be made for reservoir cooling.

A partial step towards the all electric machine is represented by a number of models, mainly in the larger sizes, which have hydraulic drive for all functions except screw rotation, which is by a servo electric motor. Again the principle is not new. Electric drives were used for screw rotation at one time but the motors were of a design now considerably outdated by servo types. The performance of a modern machine with electric screw drive is far superior and considerable savings in energy are possible.

▲ Multi colour and multi component machines.

Mouldings such as car rear lamp lenses and typewriter keys are made in two or three materials of different colours. Other mouldings may be made in two grades of material, one hard and the other soft, such as a sink plug with a rigid centre and a flexible seal around the edge. Both kinds are made on multi colour or multi component machines, 'component' in this context meaning each of the materials used, not the part produced. These machines have a separate injection unit for each colour or component. They may inject simultaneously (bi-injection) or one after the other (co-injection) according to the requirements for the moulding. There may typically be two or three injection units arranged for centre line or parting line injection (see page 161) from the side or the top. Most machines have a rotating platen for the mould so that it can be turned to each injection unit in succession, but there are also systems with which the mould stays in one position and melt flows for the different colours are directed by valves within the tool.

Machines for marbling or tortoiseshell effects (see 'Marbling units' page 79) may also have more than one injection unit, but in the mouldings the colours are not discrete but intermingled.

▲ Multi station machines.

The production of parts with thick wall sections involves prolonged cooling times and multi station machines, which provide positions for a number of moulds, are used. They permit moulds to cool without holding up the operation of the injection unit. For example, a footwear moulding machine to make thick soled Wellington boots may have 12 moulds arranged on a carousel. Material is injected into each mould in succession and the rotation of the table is sequenced. Each

boot is given a sufficient period for cooling until it arrives at the ejection station, where it is taken out of its mould. The following station is the injection station and the sequence recommences. Other types of multi station machine are used for insert moulding and permit inserts to be loaded into a mould by hand at one station during the cooling period of another. (See 'Insert moulding' pages 108, 163.) Electric plugs are made on insert machines, the material being moulded around the pins and terminals so that the plugs are integral with the ends of the cable. Injection and clamping units of multi station machines are either horizontal or vertical in a wide range of combinations. The clamping unit may have either a shuttle table for two moulds or a turntable for two or more.
The term 'outsert' might be encountered. An outsert is the same as an insert, in the sense that it is a metal article inserted in the mould, but the material is moulded inside it instead of outside, as in providing a flexible or low friction plastics lining within a metal ring.

▲ Double daylight machines.

One manufacturer produces large machines in which two moulds can be mounted in individual mould spaces one behind the other on each side of a central platen. The moulds open and close alternately and this allows one to be injected with material as the other is cooling so that maximum use is made of the injection unit. Machines of this kind can be employed to make paired mouldings such as near and offside car door panels or front and rear bumpers. The two mouldings perfectly match because the same material is used for each and injection conditions are identical.

▲ Machines for unusually large mouldings.

There is an upper limit around 300 mm to the size of screw that can be manufactured simply because the highly specialised cutting lathes on which they are made do not exist beyond a certain length and capacity. If an unusually big shot volume is required and it is not possible to make a screw large enough for it, there are two solutions: -

(i) The machine is made with a large melt accumulator. This is a cylinder into which a quantity of plasticised melt is transferred from the screw. The accumulator will be of sufficient volume for the moulding and, when filled, the material is injected into the mould by a piston.

(ii) The machine is made with several reciprocating screw injection units with an adequate aggregate volume. It may be better for these to be disposed individually around the clamping unit at 90 degrees to each other than all grouped together. Also a robot will be needed for demoulding and require access to the tool. A vertical clamping unit may therefore be more convenient than a horizontal as it provides access for the mould on all four sides. Machines for exceptionally large mouldings such as telephone kiosks are therefore made in this arrangement, which is also applicable to multi colour and multi component moulding machines that require several injection units.

▲ Rebuilt machines.

Several companies specialise in rebuilding injection moulding machines and these are often attractive alternatives to new models. Machines of good quality up to about 15 years old can be rebuilt with a complete mechanical overhaul and replacement of worn parts, a new and up to date hydraulic system and an electronic control system to customer's choice and specification.

Rebuilders mainly work to order and do not normally have reconditioned machines in stock for immediate delivery. A rebuilt machine therefore usually has to be ordered well in advance of when it will be needed. A customer can either have one of his own existing machines rebuilt or ask the rebuilder to find a suitable old one for reconditioning, which normally presents little difficulty.

Cost of a rebuilt machine may be about 40 per cent less than that of a new one of similar kN capacity, but a comparison may be difficult to make if the basic specification of a rebuilt machine, defined many years previously, has become out of date. Clamping unit dimensions, in particular, may not compare favourably with those of a new model of similar kN capacity, but if these are adequate there is much to be said for a rebuilt machine and larger sizes will probably represent good value for money.

Although at least one machinery manufacturer offers a machine renovation service, rebuilding companies are typically small engineering concerns who are able to give good after sales support for their products but, by their very nature, do not employ personnel on the development of novel technologies or advanced machine development. Moulders who require this kind of support for the introduction of new projects will need to go to one of the major manufacturers.

▲ Low shot weight machines/Manual and semi automatic machines.

Although this book is chiefly concerned with mainstream machines with shot weight capacities from about 25 g up to the largest sizes (more than 120 kg), smaller plunger injection machines with capacities down to only 4 g call for mention. It is often not practical or economic to mould extremely small parts on machines of about 200 kN clamping force and models in the 40 to 100 kN range come into their own for this class of work. They are mainly used to make technical parts such as gear wheels, washers, bushes and rings, mainly in engineering grade materials, or special medical and packaging components. Most of the machines in this class are pneumatically operated and of the plunger injection type. Machines for micro moulding to make parts weighing about 0.1 g are also available.

Machines with simple manual or semi automatic controls are manufactured. These are intended to be used by an operator and are for laboratory, prototyping or short run production work. They range from lever operated benchtop plunger injection models to hydraulic types of about 200 kN with reciprocating screw injection units. Their cost is less than for fully automatic machines as their controls are fairly basic.

chapter 6

Making a production cost analysis.
Recording machine operating costs as a guide to future machine choice.
Justifying the cost of an expensive machine.
Budgeting for mould and machine.
Sourcing moulds.

At some stage during the process of choosing a machine, and apart from considering the technical aspects involved, a production cost analysis for the part concerned should be prepared as a basic guide to machine and mould selection. Hypothetical figures are used here to illustrate this simple method: -

▲ Production cost analysis for part.

Annual requirement: 9 million

Part can be made in a four impression mould on a 500 kN machine
  Cycle time 8 seconds
  Cycles per hour 450
  Cycles per year (50 weeks x 120 hours) 6,000 x 450 = 2.7 million
Annual production capacity 2.7 million cycles x 4 impressions = 10.8 million

Production capacity therefore adequate for annual requirement. Margin for less than 100 per cent utilisation of machine allowed.

100 per cent machine utilisation throughout the year is an unattainable ideal under real conditions, but 90 per cent can be achieved in a well managed moulding shop. This means that the machine is run for 90 per cent of the total working hours available.

Machine and mould costs	Per year £	Per hour £
Machine depreciation First cost £42,000 Assumed life 10 years	4,200	0.70
Mould depreciation First cost £12,000 Assumed life 5 years	2,400	0.40
Labour (10 per cent of one operative's time)	1,500	0.25
General overheads including electricity*	6,000	1.00
Sub total	14,100	2.35

Material costs

Material: PP (polypropylene) Cost per tonne £900 Weight of part 10 g 10,800,000 parts x 10 g = 108 tonnes		
Cost of material per year 108 tonnes x £900 =	97,200	
18 kg per hr @ £0.9/kg =		16.20
Total production cost	£111,300	£18.55

Cost per part 1.03 p (Total production cost of £111,300 divided by 10,800,000 parts)

* Electricity cost for a machine, excluding for water cooling, compressed air and other applications e.g. material drying, may be estimated on the basis of approximately 0.75 to 1 unit (0.75 to 1kW/hr) per kg processed. For the above example the cost would come to approximately 12 p/hr and £650/yr at an average tariff of about 0.8 p per unit.

Several analyses may be made along these lines, taking a different size of machine and mould in each case. The required rate of output determines the size of the mould and in turn the size of the mould determines the size of the machine. The first question to decide is how many cavities, or impressions, the mould should have. Should there be one mould with a large number of cavities or several smaller moulds each with fewer cavities? In the first instance a fairly large machine will be needed and in the second several smaller machines. The more cavities a mould has, the greater the output of the machine per cycle and therefore per hour, but some moulders prefer to have the measure of flexibility that running several smaller moulds and machines allows. There are a number of software programs for these preliminary calculations.

▲ Recording machine operating costs as a guide to future machine choice.

A reason for keeping cost records for machine operation is to be able to compile information which will be useful when future machine purchasing decisions have to be made. The production cost analysis shown above is based on an extremely simple breakdown of costs but may form the basis of a recording system which could prove to be invaluable. The usefulness of such a system is greater the more headings it has, although a moulder may see little point in refining a costing system, assuming he has one, because of the heavy predominance of material costs. This, however, is a poor excuse, particularly as several easily kept PC programs are available.

In the above analysis, 'General overheads' covers many items of expenditure but in an actual costing system these should be broken down so far as possible. Water cooling costs, for example, should have a separate heading and if possible be further broken down between oil reservoir cooling and mould cooling. The equipment for oil reservoir water cooling is usually separate and different from that for mould water cooling. The running costs of each should be separately recorded and furthermore identified with individual machines. Keeping track of oil cooling costs may reveal a tendency for one type of machine to run hotter than another relative to the rate of throughput of material and this may be investigated.

Even although material costs are high and tend to make items such as energy consumption and water cooling seem insignificant, it is still prudent to monitor and analyse machine performance and operating costs. It is often possible by doing so to detect the effects of wear or other causes of deterioration at an early stage and take remedial action in good time.

One moulder who recorded the downtimes of his machines discovered that only a minority of them, all of the same make, accounted for more than half of the total time lost by all machines on account of breakdowns. It is almost impossible to assess the relative merits of different types and makes of machine until a period of actual use has elapsed and experience gained. A user should not allow this period to pass without implementing costing and other performance record systems. The chief object of cost recording is not just to keep track of costs for the sake of doing so but to compare the operating costs of one piece of equipment against another, and detect any deviations in ongoing costs from the normal levels.

Compared with manufacturers in other industrial sectors, makers of injection moulding machines are perhaps culpable in not providing much detailed data about operating costs. Buyers have a right to expect this but few customers have ever asked for such information and the makers in turn have never seen cause to oblige when not required to. In the absence of figures from the machine manufacturers, a user can have only his own experience as a guide to future machine selection.

▲ Justifying the cost of an expensive machine.

Although basically simple, the above analysis shows that the initial cost of a machine and mould has only a minor bearing upon final part cost because of the predominant influence of the price of material. This being so, there is an argument for buying an expensive machine and mould so that a high rate of output can be achieved. A greater number of parts is produced per hour or per year and the allocation of general overheads is at a lower amount per part. The cost of producing the part is considered as distinct simply from the cost of the machine. Most moulders will be aware of this and put the upgrading of their machines fairly high on their list of priorities, but the option may not be available to a company making a start from modest beginnings on a small amount of loan capital.

If it could be said that there are three classes of machine, namely low, medium and high priced with corresponding levels of quality, it may be agreed that a decision to buy a low priced model is solely based on cash limitations, in other words that a more expensive machine simply cannot be afforded despite a preference for one of better quality. Little persuasion is necessary to make anyone prefer a medium priced to a low priced machine, provided it can be afforded. The case for buying a high priced as against a medium priced machine, however, needs greater justification. Is the extra cost justifiable when the medium priced machine provides an acceptable standard of performance and reliability?

Because this is so, the suppliers of machines in the high price category advance cogent arguments to persuade buyers with no previous experience of running machines in this class to choose their products. One of these arguments is that an outlay on a machine twice as expensive as another is justifiable with only a relatively small increase in the cycling rate:

An expensive machine could most likely achieve a much greater reduction in cycle time than the 5.5 per cent quoted in the calculation on the following page. This is only the 'break even' figure at which its peformance would be at parity in cost terms with the less expensive machine.

Cost of machine	£50,000	£100,000
Hours of life over 10 years	60,000	60,000
Cost per hour	£0.83	£1.66
Amount of extra hourly profit required to justify more expensive machine		£0.83
Hourly turnover		£15.00

Increase in cycle time required to achieve £0.83 extra per hour: 5.5 per cent, e.g. down to 18.9 from 20 seconds

A further consideration in favour of high quality machinery is that it should offer better opportunity than lower priced for the minimisation of reject rates. In many plants, the injection moulding process is followed by 'downstream' operations such as printing, decorating and assembly which incur high reject rates. These are often intractable and offer no opportunity for improvement. In order to reduce the overall reject rate, those processes in the chain that do lend themselves to better control and the reduction of reject rates will be given full attention. Injection moulding with high standard machines is likely to be one of these and the potential for improvement may be fully exploited.

As an example, injection moulding may incur a reject rate of 1 per cent and a downstream process 3 per cent. The nature of the downstream process may prohibit any improvement but the reject rate of the injection moulding stage may be reduced to 0.5 per cent, which would be worth achieving. A high but unavoidable reject rate with a downstream process should not be a deterrent to trying to improve it elsewhere in the production sequence, rather the reverse.

Accurate part dimensions are often required by automated equipment for handling and assembly. For example, parts for a certain type of pen are made to an accuracy of +/- 0.04 mm to avoid jamming in the assembly equipment feed system. If a machine can consistently produce parts within tight dimensional tolerances, there will be a lower reject rate downstream when handled by equipment of this kind.

Many newcomers to injection moulding may be unaware that such fine precision can be attained by the process. Improvement of materials and the application of electronic controls to moulding machines has made it possible and greatly widened the range of components which can be produced in plastics instead of as machined metal parts.

A yardstick by which an hourly running cost calculated for one's machines may be measured is the prevailing rate charged by a trade moulder. There exists, albeit on an informal basis, a generally recognised scale of rates for the hourly use of a trade moulder's capacity according to machine size. These rates vary with pressure of work and the level of economic activity, dropping when this is low, but are otherwise fairly consistent.

A way of selecting a machine is to approach a trade moulder in the first instance with the intention of running the mould on one of his machines. In all probability the moulder will be pleased to collaborate and offer to quote for the construction of the mould as well. In due course the mould is run on one of his machines at a certain hourly rate. If the customer later decides to buy his own machine and production on the trade moulder's machine has been satisfactory, the choice of machine, or a name on a short list, is fairly obvious.

▲ Budgeting for mould and machine.

Moulds and machines should be matched in terms of their respective quality. Given that the moulds have been well designed and made, an expensive machine can be expected to yield a higher output rate than a less expensive machine, although the latter can be nevertheless return a good performance. Many intricate mouldings in demanding materials are profitably produced in the UK and on the Continent on machines of medium price, but this is mainly the case where the highest possible rates of output per hour are not demanded.

If a mould is not designed and made to the highest standards, it will not enable the expensive machine to give of its best. High grade machines enable the highest rates of output to be obtained from expensive moulds.

A number of moulders grade their machines according to their individual standards of performance and similarly assess their moulds according to the level of machine capability these demand. Machines and moulds are then appropriately matched. Moulds which do not need the superior control and processing ability of the more expensive models are run on the less expensive machines, and the capacity of these is allotted to the manufacture of the less demanding products. There are proprietary systems and equipment which enable machines to be calibrated and graded. Calibration means that machines are assessed for performance on a common basis.

Ideally mould and machine should be matched in terms of quality and, as a fundamental principle, mould quality is more important than machine quality. Mould quality is the prime factor and the foundation upon which choice of machine should be based. Planning and budgeting for a new project should therefore begin with consideration of the moulds or moulds, followed by machine selection, rather than the reverse.

A prospective buyer may have a certain budget for a new project to cover the cost of the mould and that of the machine. The following options can be considered when dividing the sum of money between the two:

- **A. Top class mould + high grade machine**
- **B. Top class mould + medium grade machine**
- **C. Medium class mould + high grade machine**
- **D. Medium class mould + medium grade machine**
- **E. Medium class mould + basic grade machine**
- **F. Basic class mould + medium grade machine**
- **G. Basic class mould + basic grade machine**

The combinations of top class mould + basic grade machine and basic class mould + high grade machine have been omitted. These are unrealistic.
Considering these alternatives in turn: -

A. This is the ultimate ideal but the most expensive. Where the highest possible rates of output, such as for mass volume packaging or medical products, are required, however, it is the only option.

B. The next best option to choose if A is beyond budget, but the rate of output cannot be expected to be as high.

C. This option is ranked lower than B because mould quality is the primary factor in mould/machine performance. Much depends upon the type of product, but in some instances there may not be a great deal of difference between the results obtainable from B and C. A top grade machine may draw the best possible performance out of a medium quality mould. Choosing between B and C may therefore be decided on the basis of which combination is the less expensive. Although a top grade machine will cost more than a medium grade, it is possible that alternative B will cost less overall than C when mould cost is taken into account. This will depend upon the types of mould and machine being considered.

A point in favour of choosing B instead of C is that option A will remain open for the future. The medium grade machine can be replaced at some time by a top grade and the most desirable combination attained. This would not be possible once a medium quality mould was purchased.

D. This is obviously a compatible match and a sensible way to spend one's money if the higher alternatives are too expensive.

E. The option to choose if D is above budget, accepting a lower rate of output.

F. This is probably not a sensible option unless it is envisaged that medium quality moulds will be obtained at some stage in the future. If E is above budget, it would probably be better to opt for G. It is not likely that the performance of a basic class mould will be sufficiently better in a medium grade than in a basic grade machine to justify the higher cost of the medium grade model. The inherent performance limitation of a basic class mould would probably frustrate the purpose of obtaining a medium grade machine. An actual trial of the mould in a machine would enable the best decision to be made.

G. As D, a compatible match and good value for money where the budget is limited. The danger is to spend so little that both mould and machine are of unworkably poor quality.

▲ Sourcing moulds.

How is a buyer who is unfamiliar with the mould and machinery market to know how to follow this suggested grading and matching process with any confidence? He can wholly relieve himself of the task and go to a machine manufacturer and ask for a quotation for a machine plus mould. These are known as 'package deals' and all machine manufacturers are accustomed to handling them. It is true that the mould will be marked up in price, but the machine manufacturer will carry out the initial trials of the new mould and ensure that it is ready for production before delivery to the customer.

If it is preferred to order the mould and machine individually and from different sources, the buyer will have to assess quality himself and this will be difficult in the absence of any relevant experience. One mould looks very much the same as another to the unitiated and so do machines, and the following general advice is intended to give some guidance on how to go about the task. Another course is to engage the services of an experienced consultant.

In the UK, and in other countries, a mouldmaker is chosen in most cases on the simple basis of how near he is to the user. In the event of a mould needing repair or modification and having to be taken back to its manufacturer, the shorter the distance involved the better. Most mouldmakers have a core business of customers within a few hours drive although some, who specialise in certain types of tool for particular product categories, will have a nationwide clientele. UK toolmakers work to high standards in good materials and it can generally be accepted that their products will be in the highest category, or at least medium grade. Prices quoted will reflect this. For moulds made in the UK, the alternatives regarding choice of machine as listed above are therefore A to E inclusive.

British moulders buy tools from abroad for any of three main reasons: -

(i) for highly specialised tooling of a type not made in Britain, or if made in Britain available only on long delivery. This type of tooling is top class and expensive.

(ii) to get a medium quality tool at a lower price and/or earlier delivery than from a UK supplier.

(iii) to buy a mould at lowest possible cost, accepting that quality will equate to the price paid.

Discussing each of these in turn:

(i) Exchange rates in relation to the £ Sterling enter into consideration here. The mould will almost certainly be in the top grade category and the mould/machine options as listed above are **A** and **B**. Purchasing abroad will almost certainly mean obtaining the tool from a country whose exchange rate is unfavourable and no price advantage will present itself.

(ii) and (iii) Moulds will be either of medium quality, in which case the alternatives are **C, D** and **E**, or of basic quality, leaving the alternatives **F** and **G**. The more obvious sources will be those countries whose exchange rates in relation to the £ Sterling are favourable to the UK buyer. This chiefly means less advanced countries in Europe and the Far East. One may buy a medium quality mould in these countries but at a lower price than in Britain simply because of a favourable exchange rate. The attendant risk is that, although there are many mouldmakers in developing countries as well equipped as any in Britain and whose products are of high quality, there are others who have not attained the same engineering standards as observed in the UK. The question is therefore how to be assured that the chosen mouldmaker can produce a mould to an acceptable standard and this is largely a matter of laying down a specification to which he has to work.

The quality and cost of a mould largely depend upon whether it is hardened. Vacuum hardening is a typical process. The mould is placed in an oven, made incandescent and quenched with liquid nitrogen. The sudden contraction of the steel that takes place densifies and hardens it, but the finishing tasks which need to be done after hardening take considerable time - simply because the metal is hard - and this increases the cost. Polishing, for example, may need to be done after hardening if the moulding has to have a smooth and glossy surface.

Most UK mouldmakers are equipped to produce hardened tools but a buyer seeking to obtain a hardened mould from abroad should assure himself that the manufacturer is experienced in producing tools of this description. A hardened tool has a potential life of, say, five years, but only 12 months will be sufficient for many products that are not intended to be marketed for longer than a couple of seasons and a less expensive tool which has not been hardened, or hardened by a less effective process, will be adequate. Whilst UK mouldmakers are quite capable of making tools of this kind, buyers often tend to look abroad for suppliers to obtain low price or early delivery.

Low cost mouldmakers, anywhere in the world, mainly compete in the field of unhardened tools, or tools which have not been subjected to one or other of the more effective hardening processes, and the low prices they offer will probably be compounded by a favourable exchange rate. The moulds they produce are usually for parts of simple shape that do not call for intricate machining, and if internal mechanisms are required these will probably not be either particularly complicated or made to extremely high precision.

At least one British company which buys its moulds from overseas has a strict specification which lays down steel quality and other technical details to which its mould suppliers have to adhere. In the absence of such a specification, the supplier himself is left to decide how he will build the mould and will tend to make it at lowest cost to himself. This will be at sacrifice of quality.

Other factors affecting the cost of a mould relate to its function to enable heat to be removed from the moulded part as quickly as possible, that is as a heat exchanger. This is only slightly less important than its task to impart shape to the molten material. The rate at which heat can be removed determines the profitability, or even non-profitability, of making the part. The efficiency of the mould in this respect depends to a great extent upon mould design and how the cooling channels are disposed in relation to the impression or impressions into which the hot melt flows. Ideally the melt should be cooled evenly, that is that no parts should solidify before others as this creates stresses in the component.

Cooling time is governed by how quickly the hottest part of the component takes to cool and the mould cooling system should be designed to extract heat from this area at a faster rate than elsewhere. This is often difficult as certain essential mechanical elements such as parts of the ejector mechanism may be in the way of where it would otherwise be desirable to run cooling channels. The problem is often overcome by using metals such as beryllium copper, which has a high rate of thermal conductivity, in the construction of the mould or special fittings which enable heat to be extracted from 'hot spots', that is areas of high concentration of heat. These will add to the cost of the tool.

chapter 7

Selecting a supplier. Distributors and agents.
Importance of first choice.
Judging personnel quality.
Drawing up a machine specification.
How manufacturers work out their machine prices.
Influence of competition.
Dependence upon suppliers.
Delivery time.
'Inhouse' and 'trade' moulders.
Late specification changes.

There are a few British manufacturers of injection moulding machines but for the great majority of makes sold in the UK there is local representation by an agent or distributor for a foreign company. A visit to the premises of the agent or distributor can be reassuring but whether one also visits the factory abroad where the machines are built is worth while only when the amount of business to be done is substantial.

▲ Distributors and agents.

Although the terms are often loosely used to mean the same, there is a difference between an agent and a distributor. An agent represents a manufacturer and obtains orders for machines but does not invoice the customer. This is done by the manufacturer, who pays a sales commission to the agent. A distributor buys machines from the manufacturer and sells these to and invoices the customers. The customers therefore directly pay the distributor and not the manufacturer.
It follows that a distributor will have a considerably greater cash flow than an agent, who does not buy machines from the manufacturer or hold a machine or spares stock (except on consignment, that is the items are part of the stock of the

manufacturer and on his books) and that a distributor is in a much bigger way of business with a considerably higher annual turnover. A buyer will therefore feel that he may have greater confidence in dealing with a distributor than an agent, and the larger the buyer the stronger this inclination is likely to be. Most big organisations prefer to buy from well founded distributors when they are spending large sums of money.

A prevalent opinion among buyers is that, because a commission agent will probably hold many agencies, he will tend to concentrate at any one moment on those that will be the most immediately lucrative to him and ignore the others. This is not wholly valid because any agent who behaved in such a manner would quickly lose his customers' confidence and go out of business.

More typical behaviour by an agent is that he has a relatively small clientele of customers to whose interests he attends assiduously and spends less time, proportionately, than does a distributor with a sizeable sales force in trying to gain business from new customers. He pays attentive service to his core customers in order to fend off competition from bigger distributors and the customers benefit from this.

There are, however, few makes of machine represented in the UK by agents as distinct from distributors. Agents either represent fairly expensive special types which sell only in small numbers or, at the other extreme, machines in the lowest price category. In either case the volume of sales in terms of value would probably not support a distributorship. Where price is a major attraction, buyers may not expect to enjoy the same standard of after sales support as a distributor can give, but they might even feel that they are getting a better bargain buying from a small agent than a large distributor.

It is also possible to buy machines from foreign manufacturers without representation in Britain, but the attendant risks hardly need to be described. This course is likely to be undertaken only by entrepreneurially led companies rather than publicly quoted concerns. Buyers are known to visit plastics exhibitions in the Far East and purchase machines, particularly in the larger sizes, off the stands of Pacific Rim manufacturers at about two thirds of European prices, but it is probably immediate delivery that is more the motive than any price advantage.

Distributors may be independent companies or, as is most often the case, partly or wholly owned by the manufacturer. Where the name of the distributor is similar to that of the manufacturer, it does not necessarily follow that the manufacturer is the parent company. Some customers may feel more reassured dealing with a distributor that is wholly or partly owned by the manufacturer. This inclination derives from the history of a some makes of machine for which distributorships have changed frequently over a period of years, with personnel changes on each occasion, before settling down with a directly owned subsidiary. In cases such as these, however, the root cause has often been that the manufacturer has gone through frequent changes of ownership and erratic distributorship arrangements have arisen because of this.

Another fear by some prospective customers is that an independent distributor is more likely to come adrift from its principal than one which is wholly or partly owned. This is generally baseless as most independent distributors have successfully represented their principals for many years and can confidently be judged on their record in this respect. Successful relationships between manufacturers and distributors depend equally upon both parties and the distributor is not always to blame in the few instances of where they break down.
Nevertheless it happens that a manufacturer seeking to break into the UK market may first appoint an independent distributor and in due course, after a customer base of viable size has been built up, set up its own subsidiary. This is, however, often achieved by the manufacturer buying out the distributor. The customer notices no difference in personnel or the broad business relationship and probably welcomes the change.

Generally, doubts that a buyer might have about dealing with a distributor which is not a subsidiary of the manufacturer are unfounded. It is the strength of the principal that is the primary factor. A distributor for a strong principal is more secure than one for a principal in whom full confidence may be reserved, and whether the distributor is an independent company or a subsidiary makes little difference.

All the major manufacturers serving the UK market may be considered to be well established and highly unlikely to disappear, and a customer need hardly worry that the maker of any machine he buys will go out of business during its operating life, leaving him stranded without spares or service. Where this has happened in the past, former personnel of the extinguished company have almost invariably set up on their own account, bought the spares stock and continued to provide service to users.

The supply picture is broadly one of stability and maturity because individual manufacturers have come together as parts of major groups over the years. Eighteen makes of Continental machine exhibited at the Interplas exhibition in London in 1973 were still available on the UK market 22 years later. This is not a bad record and may reassure potential buyers.

Other makers whose names disappeared during the intervening period had been taken over by one or other of the 18, and some makes were discontinued because the companies which produced them, large engineering concerns, withdrew from the plastics machinery market in order to concentrate upon other sectors. A few makes no longer available 22 years later had in any case failed to make much impact on the British market. These remarks apply to Continental makes. Eight British makes of 1973 no longer existed in 1995 but six others were still current, all but one of these under successor companies. A few new British makes also came into existence in the intervening period.

British made machines range from low shot weight pneumatic benchtop models to conventional hydraulic machines up to 1,700 kN clamping force, also special machines for the footwear and electrical accessory industries. Most UK manufacturers are small independent concerns.

Some customers prefer to deal with a major supplier because they consider that only the wide range of models and technical support available from such a company can meet their present and future needs. Others feel happier dealing with a smaller supplier with whom they can maintain a close relationship and whom they believe appreciates their custom more than a large organisation would. This accounts for the continued existence of certain suppliers who deliver only a relatively small number of machines a year, possibly only a dozen, compared with the larger suppliers who expect to sell at least 100, or at least 200 in the smaller ranges. Broadly, though, a supplier needs to sell a minimum of 10 to 12 machines of average size, that is from 900 to 2,000 kN, a year to remain viable and cover the running costs of the customer facilities he must provide.

Neither of these attitudes on the part of buyers can be said to be more sound than the other and the adequate supply position of injection moulding machines in the UK enables both to be accommodated. Loyalty by a buyer to one or a few makes of machine may certainly be said to simplify the machine selection process. Some moulders opt to standardise upon only one make of machine, but others believe it is wiser to have at least two so that neither supplier will be tempted to slacken his standard of after sales service. There is also a reason to buy from more than one source because technical progress between makers is not uniform. One manufacturer may make a particular advance and it will be some time before the others catch up. If a customer sticks to only one make, he may deny himself the opportunity of taking advantage of the latest developments.

Suppliers counter these arguments by saying that a user can expect better service if all his machines are of one make as they can less afford to lose the customer's business. Also, if a service engineer pays a visit, the more machines he can attend to on the same occasion, the lower the charge to the user.

No guidance on this question can be given here as the decisive factor is the attitude and judgment of the individual making the purchase, but it is true that when a customer loses confidence in a supplier it is most often because of a breakdown in personal relationships than technical shortcomings in the machines. Liability to a particular mechanical or other failure in a certain type of machine may be tolerated up to a point but never instances of personal indifference or commercial bungling by the supplier.

Importance of first choice.

A result of the need to maintain good relations with core customers is that suppliers tend to have customer bases which are narrower than may generally be supposed, and in fact suppliers often derive the greater part of their turnover from only a small number of buyers. The picture presented by the moulding industry is, and long has been, one in which many individual companies have either total or almost total affiliation to a particular manufacturer, according to size ranges. Where there is a variety of makes of machine, it is often because of discontinuity in management.

It follows that when a company is entering injection moulding for the first time, the selection of its first machine will have a strong influence upon future purchasing decisions. The purchase of the first machine therefore needs to be carefully considered. If machines of good quality are acquired in the first instance and prove to be reliable, there will probably be no reason to change to another make when additional machines are needed.

Whether a visit is to be made to the manufacturer will probably depend upon the value of the order to be placed, any need for technical discussions with experts based at the factory, and the distance involved. All major Continental makers who supply the British market are only a few hours flight away and this should present no problem, but a trip to Japan or North America would have to be more carefully considered. It is unlikely that a buyer would entertain the idea of a familiarisation visit to any manufacturer in the Pacific Rim countries as keen price is the main attraction of the products from this area and to make an expensive trip would defeat the object.

If a prospective buyer visits a manufacturer, it will probably be to inspect the way that the machines are built and judge whether this is efficient. It is, however, hardly possible for anyone but a professional production engineer to make a proper appraisal along these lines, particularly in the course of only a short trip.

The best that can be done is to form an overall impression from the layout of the factory. Is there a logical progression of work flow or does production involve shifting parts and sub assemblies long distances from one part of the plant to another? Do people, either on the factory floor or in the offices, appear to go about their work cheerfully and purposefully? Is there a tense atmosphere between superiors and subordinates or are interpersonal attitudes more relaxed? Is the company's literature well produced? Nevertheless, even if negative answers to such questions are forthcoming, it is still possible for machines of good quality and value to be produced in factories that do not make a favourable impression. Some buyers make a specific investigation into a machinery manufacturer's QC (Quality Control) procedures, and this is highly to be recommended. Comparisons may be made between different manufacturers and, obviously, those who appear to implement the most thorough methods of QC may be expected to make the better machine, particularly so far as reliability is concerned.

Judging personnel quality.

It is the usual practice for a manufacturer to test run a machine at the end of the assembly line but the length of this trial varies between 24 hours for some makers and 50 hours for others. Obviously a buyer is the more reassured the longer the test run and a question about this may be asked, although it is the more expensive and complex designs of machine with more systems to be checked that require the longer test runs.

Personal contacts made during a factory visit are probably the most valuable outcome. One may be introduced to individuals such as technical advisers with whom one may have to maintain future contact, and obviously the better the relationships made, the more successful any cooperation is likely to be. The main purpose of visiting a manufacturer's factory ought not to be to assess its production efficiency but to make personal contacts and judge the quality and helpfulness of technical and other personnel. If a good rapport is struck with individuals who will be advisers or otherwise helpful in the event of buying a machine from that factory, there is strong reason for deciding upon that particular make.

Training facilities for personnel may be inspected when visiting either a manufacturer or a distributor. These are an important part of the deal. If training can be given by the UK distributor, the cost of sending personnel abroad to a Continental manufacturer is not incurred. There is little point in investing in an expensive machine if shop floor personnel have little idea about how to operate it, or are predisposed not to accept it despite what higher management have decided upon. Acceptance by shop floor personnel is a paramount consideration in machine selection and the degree of acceptability will depend greatly upon how easily the control system can be understood. If there are good training facilities from which shop floor personnel can benefit, particularly in regard to instruction in the use of the control system, there is the stronger likelihood that friction between management and the shop floor will not arise.

A buyer might wonder how to decide if he preferred a particular make of machine but was not impressed by the distributor. Would he be better off buying a less acceptable make from a distributor he preferred? The answer is that the purchasing decision should be made considering the maker of the machine and the distributor as one entity. If, for example, a good machine is not likely to yield its most profitable performance because of poor after sales service or lack of technical support, the effect is the same as buying a less efficient machine in the first place. Broadly, however, a machine should be selected primarily on its own merits and the standard of the distributor considered as a subsidiary factor.

The main question to ask oneself about a distributor is: 'Are these the kind of people I can deal with and will understand my point of view? Will they be helpful and cooperative?' It is reassuring to find that a distributor has a staff of reasonable size and is not a 'one man band'. As well as the head of the company, there should be a competent and obvious second in command, although it has to be said that individuals in such a position have a tendency to move off and become the chief in another company as opportunity presents itself.

It is a fair question to ask how many machines the distributor has among his customers to each field service engineer. An average ratio is 250 but the location of the engineers is also relevant, particularly if one's base is not in a major industrial area of the country. The distributor may also be asked what his charges are for a service engineer's visit and how long a time may elapse before a service engineer is likely to arrive if called to a breakdown.

A prospective buyer can assess the efficiency of a distributor by the time he takes to give a quotation, whether all the questions asked in relation to the quotation are answered, and whether in fact the quotation given is for the machine asked for or something else. Guarantee terms may also be clarified and prices obtained for such fast moving spares as screw tips and heater bands. Comparisons of prices for similar spares items between different suppliers may reveal wide differences. Suppliers normally have to discount their prices for new machines fairly heavily and, to compensate for this, tend to regard sales of spares as being to a captive market and gross them up accordingly.

If a quotation appears to be low in price, considering the level of quality of the machine, it is pertinent to ask why. The supplier may have quoted low to gain the business at a loss he hopes to recoup later. This is not encouraging.

Maintaining a good relationship with a supplier is largely a matter of being open and frank, particularly if payment difficulties arise. A supplier becomes suspicious if a customer makes excessive demands, delays payments or gives poor excuses for non payment. These cloud the relationship and are ineffective anyway. The supplier will in future quote an inflated price because he no longer wants the customer, or needs some extra reward for doing business with him. In effect, this reduces the buyer's choice of suppliers. Unreasonable customers have in the past broken suppliers who bent too easily to their demands.

It is also a fair question if a supplier asks a prospective buyer, who is new to him, how much he has in mind to spend on the machine he wishes to obtain. Contrary to what might be supposed, a buyer does not sacrifice his bargaining power by divulging this at an early stage. In the real world, too, reluctance by a buyer to say how much he wishes to spend is taken by the supplier as a sure sign that he has little money and is trying to get something for next to nothing.

Another practice of which suppliers become aware arises when a customer has already decided upon the machine he wishes to buy but seeks quotations from elsewhere. This is probably to satisfy an internal company requirement that several quotations be obtained. Preparing a quotation takes time and costs money and a supplier is entitled to believe that his tender will be fairly considered and not sought solely because of the inquirer's company procedures. It is not unknown for a supplier in circumstances such as these to quote an absurdly high price to deter further inquiries from the source concerned.

▲ Drawing up a machine specification.

So far as possible, a buyer should set out his particular technical requirements and submit these in the form of a specification to one or more suppliers as a basis for quotation. Full interchangeability of moulds between machines is possible if all machines conform to a laid down specification. A newcomer to injection moulding will not have the experience to devise a long and detailed specification but should nevertheless aim to develop one for his future purchases. This is a systematic way to handle the selection process and saves time. Moreover suppliers will be impressed by such a professional approach.

The following is extracted from the specification of an experienced trade moulder. It is evident that the selection and quotation procedure is quicker when a list of requirements of this kind can be presented as the supplier knows exactly what is wanted, or is not wanted, and need waste no time making inquiries on points of detail.

Machine specification.

To Machine supplier ABC. Please quote for Model: ABC 350/1600H with 70 mm screw.
Consistency of part weight to be ± 0.1 per cent of average.
Features required:
Type of hydraulic control: Proportional.
Programmable preheating of hydraulic oil.
Vernier adjustment of screw back pressure.
Temperature control at hopper throat.
Mechanical limitation of mould opening stroke.
Clamping force control with indicator.
Core pulling for two cores with sequential control.
Independent operation of hydraulic ejectors during mould opening.
Air ejection on fixed and moving platens.
Program and monitor page for process control.
Storage for up to 60 mould programs.
Robot interface.
Socket for production record printout.

Devising a machine purchasing specification is an essential part of a systematic and planned approach to selection. The alternative is to buy purely on the basis of what happens to be available at the particular moment that a new machine needs to be obtained. Choice in such circumstances will probably be limited, with less likelihood of acquiring a machine exactly matched to requirements.

Inhouse moulders have led the way in drawing up specifications for their machines but trade moulders are increasingly adopting this practice. A specification should be reviewed and updated periodically with the addition of new features. The object is to determine and develop a standard machine specification and this is best done in collaboration with the chosen machinery suppliers.

Where two classes of machine of different makes are in use, the specifications of each should be as closely matched as possible. This is often the case in a shop in which a high grade make of machine is used for technical moulding and other 'workhorse' models for less demanding jobs. It should be possible to enable moulds to be interchanged between the two classes of machine to provide for flexibility if ever needed.

If it is the intention to expand injection moulding operations, this should be taken into account when new machines are specified. Many moulders buy machines for current needs but later find them inadequate. In particular, pump and injection unit sizes specified initially may become too small for future requirements. Most manufacturers design their ranges on the principle of modularity and provide a choice of pump and injection unit sizes for each model. The buyer can therefore specify a new machine appropriately in anticipation of future needs.

An extension of the process of devising specifications for individual machines is to make a detailed analysis of moulding operations over a period of a year or more, noting the total time each mould spends running in individual machines. An analysis of this kind can indicate the most economic range of machine sizes needed and the optimum number required in each, and this could form the basis of a replacement programme. The outcome might be, for example, to have a certain number of machines in each of only two kN sizes instead of a greater total of machines over a wider range of clamping force ratings. A discounted price could be negotiated for the supply of several machines all of the same size and specification.

Machinery manufacturers can assist in making such analyses and submitting proposals for replacement. This course is open to established moulders and those embarking on a new project if a forward production plan has been produced. In effect, the machinery supplier is brought into operations at an earlier stage than when immediate purchasing is simply being considered. If several suppliers are invited, one may show a greater degree of commitment than the others and this could augur a successful future partnership.

From the standpoint of a machinery manufacturer, international companies play a large part in purchasing decisions. Many of these are arrived at on a worldwide basis by centralised buying departments and are imposed upon local factories, even if they would independently have chosen differently. An international company's decision may, for instance, be influenced by factors concerning availability of back up facilities in distant emerging countries.

▲ How manufacturers work out their machine prices.

Several factors underlie how manufacturers arrive at the prices they ask for their machines and an understanding of these may help buyers when they come to consider competitive quotations. The basic factor is clamping force (which is also the starting point for the design). The amount of clamping force is an approximate indication of the quantity of mouldings, in terms of kg of material, that the machine can produce per hour, so far as most kinds of mould are concerned. The greater the clamping force, the larger the projected area of the moulding or mouldings that the machine can make per cycle. Therefore, other things being equal, the weight of material moulded per cycle and per hour will be greater the higher the machine's kN of clamping force.

In broad terms, a 1,000 kN machine, for example, can mould twice as much material per hour as a 500 kN. A moulder earns his income on the basis of how much material he can convert from its raw granular state into finished products per hour. A 1,000 kN machine, being able to process twice as much material per hour as a 500 kN, should therefore be able to command a price twice as great as a 500 kN.

Taking the first step, a manufacturer prices his machines according to clamping force. Note that this is not according to how much it costs him to make the machines but how much he believes buyers will, or should, be prepared to pay for them. But machine manufacturing costs are not proportional to machine size and this complicates the calculation of price.

The cost of making the control and hydraulic systems of a machine has no direct relation to clamping force. A 500 kN machine and another of 1,000 kN may have almost identical control and hydraulic systems and the cost to the manufacturer for the bought out parts will be similar. Assembly time will also be almost the same because an approximately equal number of mechanical, electronic and hydraulic parts have to be put together for each machine. There will be major differences only in the greater amount of steel used in the construction of the larger model and the time needed for machining various parts.

The manufacturer therefore needs to increase the price of the 500 kN machine to reflect its higher build cost in relation to its clamping force, but the competition from other makers has to be considered. Overall, smaller machines are less profitable to manufacture than larger models. Manufacturers of small machines tackle this problem in a number of ways. In the first instance their factories will not have heavy overhead cranes or big machine tools as needed for the manufacture of larger models. These impose high overhead costs.

It is also possible to simplify the design of small machines and omit various features which users expect to be included in larger types, and this allows prices to be trimmed to some extent. Another way of looking at this is to say that the prices which small machines can command prohibit more sophisticated features from being included in their specification. This is because their earning power is less on account of their lower throughput of material per hour.

▲ Influence of competition.

Competition between manufacturers has a great influence on machine prices. The market leader in a particular size range sets his prices according to how much he believes he can command for them and these are taken as the yardstick by competitors. In this respect the injection moulding supply sector is probably no different from any other. If the competitors with the market leader consider that their machines have certain features which make them more attractive, they will try and sell them at a higher price. If, on the other hand, they cannot match the quality and specification that the market leader offers, the only hope they have of selling their machines is at lower prices. Eventually they may be driven down to the choice of selling at all or selling at a loss, that is below manufacturing cost. Although all manufacturers have a standard price list, the prices at which they actually sell their machines is in a constant state of flux, and it is entirely right in a capitalist economy that this should be so. A number of influences are at play at the same time. For example, a manufacturer might have a customer of long standing with a large number of his machines and will almost certainly supply him at discounted prices to deter competitors. The customer, for his part, will expect this.

Prices that buyers actually pay for machines are market driven rather than being based upon manufacturing cost. The supplier gets as much as he can for his machines and the customer pays no more than he feels he should. This state of affairs is healthy because it tends to ensure that the customer gets good value for money simply because those suppliers who offer anything less will not survive. On the other hand it can be unhealthy if, to gain a particular order, a supplier cuts his price substantially. The word will surely be spread and other customers who have not been similarly treated will feel aggrieved.

The situation is particularly acute in Britain because in no other country in the world is the choice of supplier so wide. The leading dozen machinery suppliers have in aggregate a share of approximately 80 per cent of machine sales and the remainder compete for a slice of the remaining 20 per cent. In terms of numbers of machines delivered and value of sales, Germany is the biggest supplier with a share of about 40 per cent, followed by Italy with 35 per cent. The rest is probably taken up in almost equal proportions by other European manufacturers on the one hand and Japanese/Far Eastern makers on the other, with a few imports from North America.

▲ Dependence upon suppliers.

Obtaining good value for money is not entirely a matter of buying a machine which appears to be the most suitable for one's needs but also of being assured that the supplier can provide good after-sales service and a continuing supportive partnership. Selecting a suitable supplier is probably more important in regard to plastics machinery than in other fields of industry because it is a characteristic of the plastics sector that machinery users are heavily dependent upon their suppliers.

The reason is probably that the plastics industry has only a relatively short tradition compared with, for example, metalworking. It has grown rapidly in a short time and not yet accumulated an extensive store of wisdom and experience commonly shared by all those engaged in it. The situation has been aggravated during recent recessional periods because many companies reduced their skill base, that is dismissed experienced and trained personnel whose effort was not directly related to production. When in trouble, a plastics processor instinctively turns to his machinery supplier for help whereas in other industries he would attempt to struggle out of it on his own. The greater degree that a buyer, by an honest assessment, considers that he is going to need the support of his machinery supplier, the more carefully he should choose the make to buy. How this factor may be evaluated in relation to other purchasing considerations is discussed below.

Price and delivery may need to be considered even before the specifications of various machines are studied as there is little point in doing this if a number of models have to be ruled out because they are too expensive or on too long delivery. Price and delivery almost amount to twin aspects of the same consideration. Lower cost machines generally tend to be on quicker delivery than higher priced, contrary to what might be expected. One might suppose that the lowest priced machines would be in the highest demand and therefore on long delivery, with the more expensive awaiting buyers who could afford them, but this is certainly not the case with injection moulding machines.

If a buyer wants immediate delivery, he will have more lower than higher priced machines from which to choose. Higher priced machines tend to be available on longer delivery periods, which in times of normal economic activity, that is neither recession nor boom, average about 16 weeks. Suppliers of higher priced machines like to anticipate market needs and will hold some machines in stock in order to be able to give immediate delivery (expected to be chiefly required by trade moulders), but the number of machines they hold for this purpose will be small compared with what suppliers of lower priced models will stock.

Only the buyer himself can decide how much he is prepared to pay for a machine and most purchasers will determine a minimum to maximum price bracket. The minimum is fixed purely on the basis that a particular level of machine quality is desirable and is unlikely to be obtained below a certain price, and the maximum is dictated by how much the company can afford to pay.

A price bracket tends to be wider the higher that the buyer is on the price scale. A buyer with an ample budget will have a bracket in which the difference between the minimum and the maximum he is prepared to pay is perhaps 20 per cent, but a buyer seeking a low priced machine will not wish to pay a maximum price more than 5 per cent above the minimum he has budgeted for.

▲ Delivery time.

Consideration of delivery time is simple. If it is too long, the machine concerned has to be ruled out even if its price is within the minimum to maximum bracket and its specification acceptable. Buyers accept long delivery only when a machine has distinctive merits which put it well above its rivals, as is the case with highly specialised types. Delivery for these may extend to up to a year. It is prudent, if not always possible, to set a target date for a purchasing decision, that is to fix the latest date by which a choice of machine should be arrived at. The extent of choice of available makes will most likely be reduced if the decision is left until later.

The diagram on the following page depicts buying patterns for injection moulding machines according to category of purchaser. The object of including this diagram, which is entirely figurative and based on a personal overview of the UK injection moulding machine market rather than any carefully researched survey, is to enable a buyer to plot his position on it according to whether he is considering the purchase of a low, mid priced or expensive machine, and from this location to plan his machine selection process.

Buyers in area 'A' seeking inexpensive machines on short delivery are typically some general trade moulders or small business inhouse moulders. There is not a great difference between the lowest and highest prices they are prepared to pay for a machine. In area 'B', comprising most general trade and inhouse moulders, machines are in the medium price bracket but delivery periods tend to be longer. There is also a greater proportionate difference between the lowest and highest prices that buyers in area 'B' are prepared to pay. This difference is at its widest in area 'C', which concerns the most expensive machines. These are usually on long delivery. Buyers in area 'C' are specialist trade moulders and inhouse moulders with exacting machine requirements.

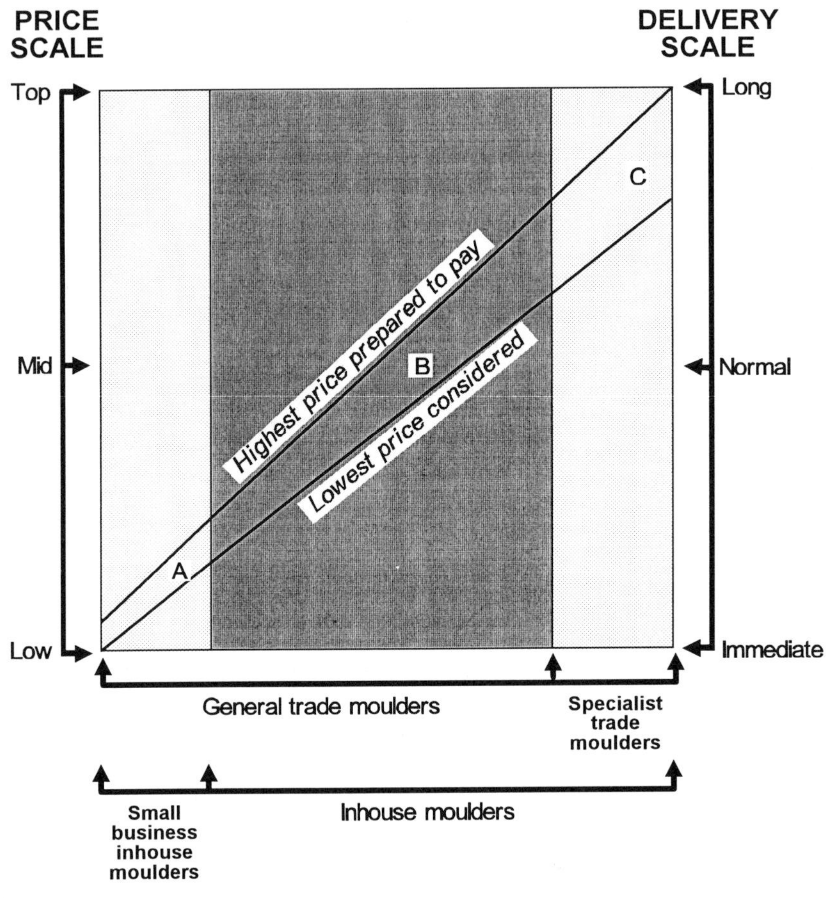

Buying patterns for injection moulding machines according to category of purchaser

If a buyer is starting a company with low capital as a trade moulder or is a small business inhouse moulder making his own product, he may plot his position on the diagram in the area marked 'A'. His choice of machine will be in the inexpensive category but there will be a fairly wide range of types from which to select, including the least expensive Continental makes. Delivery will probably be ex stock for some Continental models and others from the Pacific Rim and eastern Europe. The combination of low price with quick delivery will be fortuitous for most buyers in area 'A'. If a purchaser intentionally wants an inexpensive machine and is not concerned about quick delivery, suppliers in this category will naturally also take orders for delivery some time ahead.

If his position is in area 'B' and a medium priced machine is budgeted for, the choice of types will be about the same in terms of number of makes but the delivery situation will probably be different. The choice of machines in area 'B' comprises most Continental makes with perhaps one or two Pacific Rim types lower down towards area 'A', and some Japanese and North American higher up near area 'C'. Some makes may have to be ruled out because delivery is too long but, assuming that a fairly standard machine will be acceptable, that is without any unusual features, it is likely that a few suppliers will have a suitable machine in stock or expected shortly from the factory and can offer immediate or early delivery.

In area 'C' it is certainly incumbent upon the buyer to begin his selection process months ahead of a planned start up for the project concerned because the likelihood of being able to obtain a suitable machine ex stock or on early delivery is slender. The choice at this level will comprise a limited number of Continental, Japanese and North American makes. As indicated above, the buyer in this area should be prepared for a wider proportionate differentiation in prices between different makes than applies to low and medium-priced machines.

This does not arise from lack of competition but because the makers at this end of the market are freer to take different design approaches to the problems which confront them. These almost always turn on how to build extremely high output machines capable of producing consistently accurate mouldings - packaging products, blow moulding preforms, medical disposables and data storage disks being typical.

The result is a choice to the buyer of machines which are designed according to different concepts and principles and the wider price range arises because of this, some concepts being more expensive to realise and build than others. There may, for example, be two or three machines each specifically intended to do the same work but they will differ considerably in design. A choice between them will involve a thorough appraisal of their relative performance.

By contrast, choosing among less expensive machines turns more on non technical factors such as price and delivery. Among the lowest cost machines, chiefly designed and built to sell at a competitive price, there is much less difference in design between the various makes and in many cases it is difficult to differentiate between them in specification.

▲ 'Inhouse' and 'trade' moulders.

Moulders are described either as 'inhouse' or 'trade', although there is some overlap in that there are inhouse moulders who will undertake trade moulding to fill spare capacity and some trade moulders who have their own inhouse products for the same reason, but the classification is broadly applicable.

The work of inhouse moulders usually relates to forward projects and they have time to plan ahead. This being so, they do not normally require machines on quick delivery. Financial aspects tend to be less restrictive than in the trade moulding sector and the dearer classes of machine can be more easily afforded. Area 'B' in the diagram on page134 is therefore more relevant to them than area 'A'. Many inhouse moulders also have to meet highly demanding product standards and necessarily require machines capable of the best performance, although this is also true of some specialist trade moulders, particularly those producing high volume packaging. This requirement lifts their purchasing into the more expensive classes of machine and area 'C'.

Trade moulders undertake work for proprietary manufacturers and widely quote for this. There is heavy pressure on their capacity when there is an upturn in the economy and a greater proportion of their quotations is accepted than when business is at a low level. Much of their work in good economic times is for inhouse moulders whose capacity is overstretched. This frequently causes trade moulders to require extra machine capacity at short notice, presenting an opportunity for those suppliers able to give immediate, or almost immediate, delivery.

Other trade moulders, who work in fields less affected by fluctuations in the economy, may prefer and can afford a higher class of machine and tend to have close relations with machinery suppliers in this category. These suppliers, in turn, will try and stock machines in anticipation of their needs, knowing that if they do not they risk losing their business to suppliers of more readily available machines, even if these are of lower specification.

However, if market demand rises rapidly and unexpectedly, a supplier of higher class machines may quickly sell out his stock and have to disappoint established trade moulder customers. They will either stay loyal and await a long delivery or buy elsewhere. It is tempting for a moulder to behave unfairly in such circumstances and accept a machine from a lower priced supplier ostensibly on trial for a few months. Actually he places an order for the machine he truly wants and uses the low priced machine while he awaits delivery of the machine he prefers. When this arrives, he returns the 'trial' machine with some transparent excuse such as: 'We really want a bigger machine than this.'

This is unreasonable and the moulder should be prepared to pay a hire charge in such circumstances. Moulders who behave like this acquire a poor reputation among suppliers but because of the degree of competition there is usually a supplier who will make machines available to a customer on this chancy basis. Occasionally the machine may be retained by the customer if he finds, unpredictably, that he needs its capacity for a certain grade of work. These are the typical circumstances under which suppliers of low priced machines operate and endeavour to sell. They accept the business risks involved and their optimism sustains them, but generally their operations and customer relationships tend to be more opportunist than applies to suppliers of more expensive machines.

Neverthless they constitute an ever present threat to suppliers of more expensive machines who, in turn, cultivate and maintain close relationships with their most important core customers in order to deter competitors.

This tends to polarise the supplier/user situation to a great extent. Suppliers concentrate their sales efforts upon users who are already customers or others to whom they think they are highly likely to sell. Many suppliers serving the 'B' and 'C' areas in the diagram above, whilst they may have a large number of users of their machines overall because they have been in the market for many years, have an annual order intake arising from only a relatively few customers. These will probably be large organisations and groups whose orders greatly exceed the volume from smaller moulders.

▲ Late specification changes.

It is often the case that, once an order has been placed for a machine, there are changes in the design of the end product which necessitate alterations to the tool and, in turn, changes in the specification of the machine. It may become necessary, for example, to amend ejection or core pulling arrangements. Highly computerised manufacturing methods, as employed by most machinery manufacturers, do not allow such changes to be made without the likelihood of delay in delivery. What might seem to be a simple request will involve considerable amendments to the production schedule with a 'knock on' effect upon other machines in various stages of build at the same time.

The possibility of any such changes should be raised with the supplier when the order is placed so that any alterations to agreed specification may be anticipated and delay reduced. For example, a manufacturer may prearrange to do the necessary work to a completed machine in the service department, or it may be undertaken by the distributor at his premises before delivery to the customer.

chapter 8

Conducting a machine trial.
Checking for output capability.
Weight and dimensional consistency.
Using an analyser.
Colour dispersion.
Electricity consumption.
Decibel level.
Access for maintenance.
Purging out.
Mould safety sensing.
Mould face parellelity.
Delivery checks.

▲ Checking for output capability.

A machine trial may be carried out at the premises of a distributor or the manufacturer. The purpose is to determine that the machine will be capable of producing acceptable mouldings at a profitable rate. The following is a description of a fairly exhaustive trial and a buyer may adopt all or only parts of it: equally he may add further tests of his own. The extent to which the buyer will check the performance of the machine will depend upon his production requirements and the demands of end customers.

Material for a trial is normally supplied by the customer and should obviously be of the same kind as that intended to be used in production. Melt quality should first be examined. The usual method is to retract the injection unit and inject an amount of material slowly into a 1 litre bowl made of high temperature plastics such as polycarbonate and stir this with a heavy glass rod. Lumpy inconsistent melt is immediately obvious and the settings of the machine may be altered so that a good homogeneous melt is produced. The temperature of the sample may be taken with a probe type thermometer. As a secondary check, a weak masterbatch addition may be run with natural or crystal (glasslike) material to check the ability of the screw to achieve good melt homogeneity. Thorough colour dispersion indicates the ability of the particular screw design to produce a melt of even viscosity and temperature.

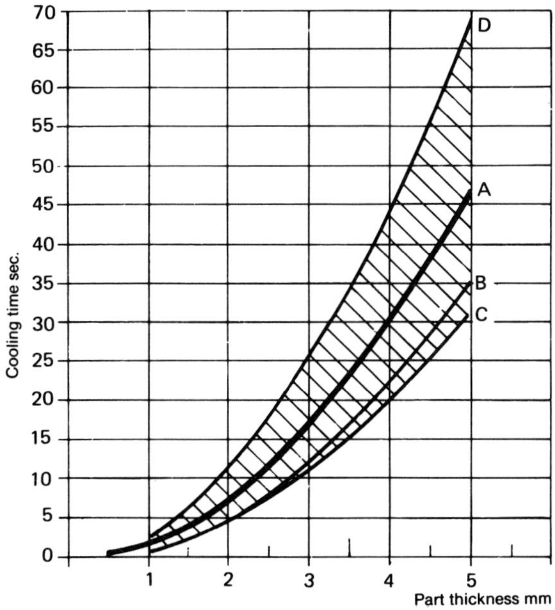

		Temperature of molten plastic	Mould temperature	Ejection temperature of moulded part
Moulding conditions	A	300°C	100°C	125°C
	B	300°C	80°C	125°C
	C	300°C	80°C	130°C
	D	300°C	120°C	130°C

Mould cooling conditions should be similar when comparative tests of different machines are made. In particular, the temperature of the mouldings when ejected should be the same and this will involve running with similar melt and mould temperatures. These graphs show the cooling times according to wall thickness for mouldings in Bayer Makrolon polycarbonate material for different mould temperatures. Graph 'A' shows the temperature of a part with 5 mm wall thickness, for example, to be 125 degrees C after approximately 47 seconds cooling time in a mould maintained at 100 degrees C. By reducing the mould temperature to 80 degrees C, as in graph 'B', the same part temperature on ejection was achieved after only 35 seconds cooling time. Mould temperatures should therefore be standardised upon when when comparing the cycle times achievable by different machines.
When mould temperature is raised from 80 to 120 degrees C, as between graphs 'C' and 'D', the higher temperature necessitates an increase in cooling time from about 32 to 68 seconds so that the same part temperature on ejection is obtained.

It is usual, in order to save time, for the mould to be delivered in advance of the trial and in this case the distributor or manufacturer will probably have made a preliminary run to determine the settings for the initial injection and holding pressure phases. However, it is wise for the customer to check that adequate cooling time has been allowed within the overall cycle time.

There may be the temptation for the demonstrator to curtail cooling and therefore the overall cycle time, and the output rate may appear to be satisfactory. But if cooling time is too short, mouldings will chill on ejection and sudden exposure to ambient air temperature and noticeably distort. As discussed before, this aspect of moulding is critical. Adequate cooling time is particularly important for presentation mouldings, that is mouldings which when in use are on view as distinct from being fitted inside a piece of mechanism or an enclosure of some kind and are unseen. A TV cabinet or the outer housing of a domestic appliance such as a food blender is an example of a presentation moulding. Distorted mouldings don't look very good. The parts should be cooled down within the mould to a certain temperature before it opens and they are ejected, and this temperature can be found by reference to the material manufacturer's literature. A contact type probe thermometer may be used to check part temperature on ejection and cooling time adjusted as required to ensure that this is as specified. The proper cycle time for the production of adequately cooled mouldings is thus arrived at, but it will also be necessary to check the temperature to which the mould temperature controller is set. Only by running tests on different machines and applying similar temperatures can comparable results be obtained. The temperature to which the controller is set should again be according to the recommendation of the material manufacturer. Wall thickness and not part weight determines cooling time in most instances. For a certain wall thickness, a heavy part takes the same time to cool as a lighter one, assuming that a mould temperature controller with an adequately powerful water pump is used.

Other checks to make before cycling the machine include an inspection for surface quality. Is the surface finish even or patchy? Is the gloss up to the standard required? Are weld lines apparent? Weld lines are caused when a flow of melt in a cavity divides to go around a feature representing an obstacle and rejoins on the other side. Where the two faces rejoin there may not be a desirable meld of the material and this is apparent as a faint line on the surface of the moulding.

Seen under a microscope it is a distinct cavernous gap and a source of structural weakness. An adjustment to melt temperature and injection pressure should enable weld lines to be eliminated as a prerequisite to a mould trial intended to simulate actual production. There is a specific book on recognising surface defects (see 'Further reading' page 175).

Hourly production rate can be calculated from the cycle time and it can soon be found whether the rate required is attainable by the machine. According to the availability of the machine and the amount of material it is desired to use in the course of the trial, the machine can continue to be cycled for as long as considered necessary. Some customers insist on at least 24 hours so that the effects of changes in ambient temperature throughout the day and night are made apparent.

Screw slip may be observed during plasticising and an adjustment to the back pressure setting may be made to eliminate it as it is undesirable. On the front of the injection unit is a horizontal slot with a scale and a moving pointer. The pointer travels backwards with the screw and it can easily be seen whether it slips or moves at a consistent rate. Variations in the temperature and viscosity of the melt will exacerbate any tendency for the screw to slip during this phase and may indicate poor screw design.

During a fairly long period of moulding, the temperature of the oil reservoir may be observed from time to time: 45 to 48°C is the accepted oil temperature range. If a high oil temperature persists, this may indicate either inadequate cooling or a poorly designed hydraulic circuit. Running a machine at too high an oil temperature is likely in time to lead to irregular operation and liability to breakdown because of seal failure. The indication of the oil reservoir thermometer should always be below the red zone by an adequate margin and stable. The temperature shown is that of the oil in the reservoir. The oil in various working parts of the hydraulic circuit such as the valves and cylinders will be at a higher temperature and should not exceed a critical level. The temperature of these parts may be judged simply by placing one's hand on them (carefully). They should not become dangerously hot.

▲ Weight and dimensional consistency.

Weight and dimensions of mouldings produced should be as consistent as possible. These two criteria are distinct but related. The end customer, who may have to assemble the mouldings with other parts, will be concerned about dimensional consistency, that is that the parts are neither too long nor too short to fit properly with others, and will specify his requirements in this respect. In particular, diameters of any holes will probably be specified to tight limits. Weight consistency is, however, a precondition of dimensional consistency and needs to be checked in the course of a trial. Moulding with good weight consistency means that few parts either over or under desired weight limits, and which therefore must be rejected, are produced. In turn this means that more mouldings per kg of material are made than when consistency is poor and part of the machine running time is wasted producing rejects.

An electronic balance is used to check weight consistency. A quantity of mouldings is collected and allowed to cool for about 30 minutes because the currents of warm air rising from hot mouldings create a turbulence that disturbs the pan of the weigh scale, which is extremely sensitive.

The mouldings are individually weighed. If a computer weigher is used, as is usually available in a manufacturer's demonstration bay, the results can be analysed and an SD (Standard Deviation) figure produced. Alternatively a pocket scientific calculator with an SD program may be used. A common requirement by machinery buyers, reflecting in turn the specifications of their customers, is that the weights should all be within 6 x SD, but it may be more stringent. The formula for the calculation of an SD figure moderates the effect of random extreme weights when the rest are within a close tolerance. The lower the SD figure, the better the repeatability of the machine. SD figures for different machines can be compared. For a full explanation of this term, the reader is referred to a handbook on SPC (Statistical Process Control) (see 'Further reading' page 175).

In the absence of computation by SPC methods, the difference between the heaviest and lightest mouldings produced may be calculated as a percentage of the overall average. The manufacturer should be asked what his machine can be expected to achieve in this respect and his claim verified, although a number of factors other than machine performance have an influence. If he confidently claims that his machine can produce a number of mouldings to within 0.1 per cent of the average weight, as may be the case with a good class of machine, it can be determined whether this is in fact attainable.

The machine should not, however, be rejected if it does not attain the claimed standard because, as mentioned, a number of other factors have to be taken into account. The percentage figure obtained can be compared with the performance of other machines. Generally, however, a modern machine should be capable of producing mouldings within a 0.2 per cent weight variation, even without closed loop injection control.

Dimensions, as distinct from weights, should be measured after the mouldings have been allowed at least 24 hours to cool. An amount of shrinkage occurs simply because of the physical contraction with reduction in temperature, but a marked further contraction after moulding takes place with certain materials because of a change in their crystalline molecular structure. The materials concerned here include PE, PP, POM (acetal), PA (nylon) and PBT. As an example of this effect, headlamp reflectors may be moulded to specific dimensions to fit an assembly. They may be perfect immediately after moulding but will not fit when needed to be assembled a few days later. This delayed post moulding contraction has to be taken into account when the mould is designed and made, but poor screw performance can vitiate the benefit of accurate mouldmaking. As previously explained, poor plasticisation results in the production of an inhomogeneous melt. In turn this is the cause of stresses in the moulding which can be so great as to distort it out of its proper shape and dimensions.

Good plasticising performance becomes highly important if consistent dimensional accuracy is critical, and the batch of parts produced during a trial should be measured for reassurance that the machine is capable of making acceptable mouldings with a low reject rate. Measurements may be taken between pairs of points as specified by the end customer or otherwise chosen and an electronic micrometer used. The mechanical thumbwheel type of instrument which involves reading off an etched scale is unlikely to enable sufficiently fine measurements to be taken.

As with weighing, input into a computer or a calculator will permit an SD figure to be reckoned but otherwise the difference between the longest and shortest parts may be calculated as a percentage of the average. This figure will not be the same as the respective percentage for weight variation but considerably lower. The following figures derived from an actual trial illustrate this: -

Analysis of weights and lengths of parts produced.

Heaviest part	85.36g	Longest part	343.24 mm
Lightest part	85.30g	Shortest part	343.189 mm
Difference	0.06g	Difference	0.051 mm
Average weight	85.33g	Average length	343.22 mm
Difference as % of average	0.07%	Difference as % of average	0.015 %
Standard Deviation	0.16g	Standard Deviation	0.014 mm
6 x SD	0.96g	6 x SD	0.084 mm

(Figures from article in *Plastics & Rubber Weekly* 'Machine on test' series published March 1, 1989. Machine: Krauss Maffei KM150-620B.)

Dimensional stability is believed to depend considerably upon the temperature of a part at the moment of ejection compared with ambient temperature. The part is suddenly exposed to a chill when the mould opens and the degree of chilling depends upon the difference between the temperature down to which it has been cooled and the ambient level.

If comparative tests are being made with several machines of different makes, each manufacturer will seek to show that his machine is faster than the others. A probe type thermometer is a valuable tool in such investigations so that the temperature of the parts can be measured on ejection. The temperature should be the same for all tests because cycle time can easily and deceptively be shortened simply by ejecting the parts at a higher temperature than applied to a previous trial.

Temperature readings for a part should always be made at the same point on its surface. This is because the surface temperature of the mould is unlikely to be consistent from one area to another. Cooling is inconsistent because the water channels are at varying distances from the mould surface and there will be corresponding differences in the temperature of the part at different points.

▲ Using an analyser.

It is possible to compare the standards of performance of different machines by means of a portable machine analyser. This is a piece of test equipment which is connected to the machine and monitors the various times, stroke distances, pressures and temperatures. One maker of equipment of this type has conducted extensive tests in the USA and devised a table by which machines may be classified according to their standard of performance. Machines in the top classification will, for example, hold overall cycle times to within +/- 0.2 second. Over a morning's trial, a new machine should be found to fall into either Class 1 or Class 2 and the table below shows some of the tolerances which should be met. It may be used as a guide or the user can impose the tolerances that he requires.

Table: Permissible tolerances for various parameters as +/- from the mean for the purpose of establishing upper and lower specification limits (USL and LSL). Figures shown are extracted from the Hunkar classification system. In its full version this contains data for nine classes of machine and more parameters than quoted here.

	MACHINE	
	Class 1	Class 2
Cycle time	0.2 sec	0.24 sec
Overall injection time	0.04 sec	0.05 sec
Holding pressure time	0.02 sec	0.02 sec
Plasticising time	0.15 sec	0.18 sec
Peak injection pressure	1.4 bar	1.6 bar
Holding pressure	0.3 bar	0.3 bar
Screw back pressure	0.3 bar	0.4 bar
Injection unit zone temperatures	1.11°C	1.33°C
Injection stroke	1.27 mm	1.52 mm
Oil temperature	1.67°C	2.0°C

▲ Colour dispersion.

Moulding may be done with material which is already supplied in a colour or in what is called 'natural' material of a neutral, translucent appearance to which a masterbatch colourant is added in small proportions, usually about 1 to 4 per cent. Masterbatch is granular plastics material which has an extremely high colour concentration. A buyer may wish to check the effectiveness with which the colourant is dispersed through the natural material and add the desired dose of masterbatch to material during a trial.

In practice, the concentration of most masterbatch colourants is so intense that in the percentage range quoted even the poorest of screw designs should be able to disperse it effectively so that full colour saturation is achieved. Additions of masterbatch in lower proportions will better reveal the colour dispersion efficiency of the screw and tests may be made with a dose of only 0.25 per cent or even less.

These will show up the difference in performance between a screw with good colour dispersion and another which mixes less thoroughly. Screws specially designed for good colour dispersion may be evaluated with weak additions of masterbatch. Colourant is expensive and may be saved if lower than usual percentage additions to natural material can be used.

▲ Electricity consumption.

If the necessary equipment is available, the electricity consumption of the machine may be measured and related to the amount of material processed. This is known as the specific energy consumption and is expressed in kW/hr/kg. In Britain what is known as a 'unit' of electricity is 1 kW/hr and, from the tariff of the electricity supplier, it should be possible to calculate the cost of energy per kg of material moulded.

With rising costs of electricity, this may be far higher than assumed and bear upon profitability if margins are tight.

The rate of electricity consumption is useful to know not only for the calculation of part costs but also as an indication of the efficiency of the drive system of the machine, although, as mentioned above, the proportion of available shot volume used is an important factor and may need to be taken into account when making comparisons between different models.

It is not necessarily the case that the specific energy consumption of an expensive machine is less than that of one lower in price. A machine with a comparatively simple hydraulic system can use less electricity than a more expensive model with many valves each of which represents a restriction to the oil flow.

Consumption may be measured for the hydraulic pump plus heaters and for the pump only. It may be possible for the demonstrator to produce graphs, using the printer port of the machine's control system, to depict the power draw over the cycle.

A clearer picture of the draw by the pump is given if this is shown without the superimposition of the draw by the heaters. Consumption by the heaters is usually about 25 per cent of the total but depends to some extent upon the flow path length to wall thickness ratio of the moulding, a high ratio imposing greater power consumption because of the higher injection pressure required.

The plasticising phase accounts for a substantial amount of the total energy consumption over the cycle and can be assessed from the graph, which will also show the peak draws of power which occur during the build up of clamping pressure and injection.

If peaks for a number of machines running in a factory randomly coincide, the maximum permitted by the electricity supplier may be exceeded.

Most moulders are not greatly concerned with electricity consumption as such, considering that if a reduction in reject rates can be achieved, this is tantamount to a reduction in units consumed. Nevertheless, expenditure on electricity can be considerably underestimated and it is good management to know the specific figure for energy cost.

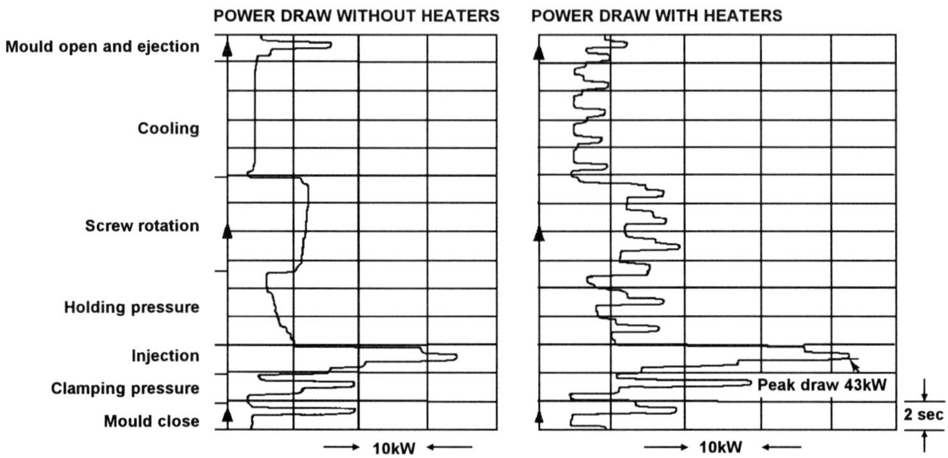

Electricity consumption of a machine should be measured during trial runs and a print out made for the power draw over a number of cycles. It is also desirable for the power draw to be traced with and without the heater bands as a clearer picture of the latter condition is given. Heater band consumption is normally about 20 to 30% of the total.

These diagrams are taken from an actual machine test and should be read from bottom to top to follow the cycle phases. They show how the power draw rises and falls according to the functions being performed by the machine. The amount of power consumed in terms of kW/hr (units) can be related to the weight of material processed.

There may be a considerable difference between the rates of consumption by different machines and this could affect profitability. Injection, maintaining holding pressure and screw rotation account for the greater part of power consumption and will vary mainly according to wall thickness and the viscosity of the material used.

▲ Decibel level.

Factory noise is becoming a major health and safety factor and a buyer may wish to measure the noise level of the machine with a decibel meter. It is frequently the case, however, that other noise sources in the demonstration area, a ventilation fan for example, will frustrate this.
Otherwise, the maximum decibel level measured from several points around the machine and 1 metre away from it should be less than 78 dB.

The maximum level is likely to be reached during plasticisation, when the screw drive motor is running, but there may be momentary pressure 'kicks' in the hydraulic circuit which exceed this. Even if noise levels are not an important consideration, quiet running is an indication of good hydraulic circuit design and energy efficiency.

Most modern machines have stylish all-enclosing guarding which looks smart and attractive but may have been designed with no regard to acoustic characteristics. The panelling, particularly around the injection unit, may actually amplify the noise from the main pump and the screw drive motor. Large flat areas of guarding panelling resonate with the pump and motor vibration and set up additional noise emission of their own.

▲ Access for maintenance.

Ease of access to filters and other parts of the machine which require periodic attention may be checked. In particular the nozzle may need fairly frequent cleaning and the injection carriage will need to be retracted to gain access to it. Too slow a retraction speed can prolong this task.

The shut off valve at the end of the screw may also need to be cleaned fairly frequently, particularly where materials "housekeeping" is poor, and this involves partial disassembly of the front of the injection unit. Some machines have better provision for access to the screw shut off valve than others.

▲ Purging out.

Any material remaining in the barrel and nozzle of a machine should be removed when a change of colour or material is made (although sometimes the necessity for this can be avoided, as when changing from a light colour to a dark). The usual way of doing this, unless a special purging agent is used, is to put some neutral material through the machine until no traces of the old remain.

Moulders who frequently change colours or materials like to check how easily this can be done on a prospective new machine and will run appropriate tests. However, the chief factor militating against quick and easy purging is the existence of a low temperature zone in the material flow path and this can be easily corrected by heater adjustment. Screws with complex flight patterns, for example, should not present any difficulty on this account and neither ought the screw shut off valve.

As an approximate guide, it should be possible to purge out fully with a quantity of material equivalent to four times the barrel inventory of the machine or eight times its maximum shot volume. The inventory is the total amount of material in the barrel comprising the volume at the front, ahead of the screw, and that around the screw in between the flights. This advice applies to a screw with a length/diameter ratio of 20:1. A screw with a higher L/D ratio would require proportionately more material. It is recommended that any screw back pressure be cancelled before purging to ensure that the screw fully returns to its back stop.

▲ Mould safety sensing.

Various design and operating features may be examined after checking the performance of the machine. All machines incorporate mould safety sensing. When the mould closes, the start of the movement is at high speed, but the final part, just as the two halves are coming together, is at low speed. The purpose is so that the expensive mould is not damaged, as it would be if the two halves were brought together quickly.

The pressure and speed for the final part of the movement can be set at the required levels. These will depend upon the configuration of the mould, a complex mould needing to be closed relatively slowly. If a foreign body should be present between the two faces, the moving mould half will 'sense' this and, because it is travelling at low pressure, come to a halt because the pressure set is insufficient to overcome the resistance of the impediment. After a timed short period during which it is stationary, the moving platen retracts and an alarm sounds or illuminates.

This feature is known as 'low pressure mould safety sensing' and prevents any foreign body which might come between the two mould faces from being crushed between them. This could damage the mould, which would be expensive to repair, as well as cause loss of production. The most likely reason for the activation of the sensing system is failure for a part to eject on the previous cycle and be retained in the mould, a condition known as 'hanging up'. Small fragments of the mould may also break off because of wear and stick to the mould face. Damage to the mould could be severe if a mould closes and is clamped with full force on hard particles of this kind.

A check on the sensitivity of the mould safety sensing feature may be made. Ask for it to be adjusted down to its most sensitive limit, that is at the lowest pressure. The manufacturer may claim, for example, that this will enable an object measuring only 0.07 mm to be sensed and this can be verified. A more simple test is to stick a postage stamp on the fixed half of the mould. The moving half should sense this obstacle, halt and retract.

Time for the trial may not allow for a mould safety sensing check but it is valuable to make it once the machine has been delivered. The machine is normally set so that the changeover from high speed to slow safety sensing during the mould close movement occurs when the mould faces are a certain distance apart. The less this distance is, the more time that is saved on the cycle.

However, the consistency of some machines may be poorer than others and the changeover may not always occur precisely at the desired point. Sometimes it may occur in advance of it and at other times overshoot. The danger is that if the moving platen overshoots the changeover point by too much the mould faces will be brought together at high speed and be damaged.

Machines are therefore set by the user so that the changeover point is sufficiently in advance of the meeting of the mould faces and the worst amount of overshoot likely to occur will not override and cancel safety sensing, but it is obviously as well to discover the machine's capability in this respect. If changeover to safety sensing always occurs at the same point on the closing stroke to within 0.1 to 0.3 mm, this can be allowed for, but if it is in terms of full millimetres and unpredictably random the risk of mould damage is much greater. A machine with a well designed hydraulic system should behave consistently and invariably make the final closing at low speed, although wear after a period of use may cause its behaviour to deteriorate.

A sign of a well designed machine is the smoothness of its mould opening and closing movements. Just as the mould closes quickly at first and then slowly in the final phase, it opens slowly, then accelerates and slows down again towards the end of its stroke. On older machines these phases are perceptibly discontinuous and make the movement look jerky, but on one of modern design they flow, or should flow, into each other smoothly and travel over the whole stroke distance appears to be one uninterrupted movement. Improvements in hydraulic valve design and control have enabled this to be achieved and contribute to energy economy and machine longevity.

Mould opening begins slowly to allow time for the moulding to become unstuck from one half of the mould. This is usually the fixed half, with the moulding travelling with the moving half, although the opposite sometimes applies. If the mould opens too quickly, the moulding might be torn apart through being pulled out by the moving half of the tool while it is still sticking to the fixed half. Separation may be assisted by blowing compressed air out of minute holes in the mould against the surface of the moulding to break its adhesion to it.

Once this phase is completed, the mould travels at higher speed to save cycle time and then slows down at the end of its stroke so as not to come to a halt with a slam. The weight of the mould half adds to that of the platen and the hydraulic system has to be capable of dealing reliably with the repeated starting, stopping, acceleration and deceleration of a heavy mass of metal and consistently keep to its preset phase times.

▲ Mould face parallelity.

Another desirable check concerns what is called the parallelity of the mould faces. With the mould faces closed and touching, but before the application of clamping force, the distance between the platens at the four corners is measured. These distances should all be the same to within a certain tolerance and the manufacturer should be able to state what this is. Uneven mould wear is likely if the mould faces are out of parallel by more than a certain tolerance and clamping force will be unequally applied.

This check should always be made when the machine is installed. Perfect levelling of the base is a prerequisite of platen face parallelity. Any twist of the frame because the floor is not flat will upset the alignment of the clamping unit. The individual mounting pads are adjusted so that the top of the machine base is level. Some moulders periodically check this in case of ground subsidence because a change of only a few millimetres can be critical.

▲ Delivery checks.

A trial as performed before placing the order should be made upon delivery to check that the machine actually supplied is as capable as the one previously tested. It is possible for slight damage to be done to the machine by jolting in transit, particularly to the control system, and this can affect performance. Mouldings should be weighed and their dimensions checked during the first two months of service.

chapter 9

A method of quantifying to make a final choice of machine.

Replacement period, depreciation and residual value.

After going through a methodical selection process, the prospective buyer may be left with considerations which appear to be unquantifiable. Assuming that several machines are acceptable on grounds of price and delivery and have been short listed, making a final choice may appear difficult and, in a company in which several people contribute to the purchasing decision, give rise to disagreements among individuals which would delay arriving at a conclusion.
The solution is that the apparently unquantifiable factors simply have to be quantified and any objection to this on the ground that they are all equal in importance rejected. An effort has to be made to evaluate each individual factor and weight it according to its relative importance. A logical approach based on systematic weighting will enable a decision to be arrived at sensibly and for it to be supported it with considered figures rather than nebulous argument.
It is necessary to simplify this system for purposes of description here. Let us imagine that the short listed machines are to be assessed on these three chosen criteria:-

- **Control system acceptability**
- **Standard of supplier's after sales service and spares back up**
- **Machine footprint**

These are taken as examples only. Anyone wishing to use this system can make up his own list, which may be longer and need not necessarily include any of these factors.

The prospective buyer gives each criterion a number of points out of 10 to reflect the importance that it represents to him. But the first, whatever it is, is given 6

points as a benchmark by which the others can be weighted. Here our first consideration is control system acceptability, about which it is unnecessary to go into detail except to say that most buyers will need to consider this factor when they purchase a new machine. Some machine control systems are more 'user friendly', that is easier to understand and operate, than others and therefore more acceptable to factory personnel. As the first on the list here, it is given 6 points.
The next criterion is the standard of the supplier's after sales service and spares back up. This has to be rated in importance in relation to control system acceptability - not an easy question but one which may be solved by considering extremes.
Which of these two alternatives is preferable? -

- To have a machine incorporating a highly acceptable control system but for which after sales back up is poor,

or -

- To have a machine with a control system that is difficult to understand and use but for which the after sales back up is good.

Put in these terms, the question suggests it own answer, but this will depend on individual circumstances. In a shop in which a wide range of products is made and machine controls are frequently altered, control system acceptability may be more highly rated than in another where the product range is not as great. Similarly, moulders with their own engineering personnel or who have local freelance service engineers readily available will not be as heavily reliant upon the supplier's after sales service organisation as others.
Let us assume that, in our example, the buyer (likely to be in the majority) has a heavy dependence upon the supplier's after sales organisation and gives this consideration 7 points, rating this higher in importance to him than control system acceptability. Note that he does not give it the same number of points. The intention in applying this system is to make distinctions in values that will enable a clear result to be arrived at. This is frustrated if two or more criteria are awarded the same number of points.
The third criterion, footprint, is taken here to cover a little more than simply the length and width of the machine but also its shape and compatibility with the arrangement of ancillary equipment around it. Some machines are compact and neatly flush back and front, have three way access to the mould space and no potentially dangerous projecting parts. Others, whilst their performance is good, may be lacking in these respects.
A good footprint may be advantageous to some moulders, for example where belt conveyor take off systems have to be provided around the machines, but in our example we will imagine that it has been rated as rather less important than control system acceptability and given 4 points.

Our criteria are thus 'scored' out of 10 as follows: -

Control system acceptability	6
After-sales service	7
Footprint	4

Note that the criteria have been assessed in isolation from the merits of the short listed machines. These are next evaluated in the following steps: -

1. A number of points out of 10 is awarded to each machine according to how it rates against each criterion. These can be called the 'machine points'.

2. The criterion points and corresponding machine points are multiplied together.

3. The figures arrived at in this way are then added to give a total for each machine and an order of choice emerges according to the scores.

To go through this part of the process, set out a table for the first machine like this: -

		MACHINE A		
Control System	6	6	36	(6x6)
After sales service	7	6	42	(7x6)
Footprint	4	6	24	(4x6)
Total			102	

Here the criteria and their respective points are listed on the left. Just as 'Control system' is given 6 points as a benchmark, Machine A is awarded six points against each of the three criteria as benchmarks against which points for the other machines will be given. A total score for Machine A is next calculated. This is obtained by multiplying the criterion points and machine points together and adding the products: the total is 102.

The other two machines are next assessed. Machine B is considered to have better control system acceptability than Machine A and Machine C even more so. Machine B is therefore given 7 points and Machine C 8 points.

The supplier of Machine B is believed not to give as good after sales service as for Machine A and is awarded only 5 points. Service for Machine C, however, is thought to be better and given 7 points. One of the attractions of Machine B is a particularly neat and compact footprint and it is awarded a generous 9 points. Machine C, on the other hand, is less highly rated and given only 5.

The table is therefore completed thus:

		MACHINE A		MACHINE B		MACHINE C	
Control system	6	6	36	7	42	8	48
After sales service	7	6	42	5	35	7	49
Footprint	4	6	24	9	36	5	20
Total			102		113		117

Machine C with a total of 117 points is thus the one to prefer. The difference between B and C, however, is so close that B may be preferred if prices are still being negotiated with the respective suppliers and B reveals a distinct advantage in this respect. Machine A is a relatively poor third choice and only a fairly substantial price discount might justify its purchase against either B or C. But if the matter of price has already been settled and is no longer an influencing factor, Machine C can be chosen with reassurance and, if ever necessary, the figures for this analysis produced to justify this decision.

Let us see how another potential buyer in different circumstances might apply this system. Compared with the moulder in our first example, this purchaser has greater self sufficiency in respect of inhouse engineering facilities and therefore less reliance upon the after sales service of the supplier. He therefore rates the consideration of after sales service at 4 points. (He has already given 6 points to control system acceptability as a benchmark.) Because he has limited factory floor space and also wishes to implement a high degree of automation, he awards footprint 8 points. The three machines are judged similarly in relation to the criteria as in the first example and the end result is like this: -

		MACHINE A		MACHINE B		MACHINE C	
Control system	6	6	36	7	42	8	48
After sales service	4	6	24	5	20	7	28
Footprint	8	6	48	9	72	5	40
Total			108		134		116

Machine B is the clear choice for this buyer, using the same set of criteria. The result would probably be different again if the set of criteria were more extensive, but the user may define these according to his own particular needs.

If there is a criterion which is exclusive to one machine, it is given points according to its relative value and the machine concerned awarded 6 benchmark points. The other machines are awarded no points because they do not have the feature.

A pivoted injection unit may be taken as an example for purposes of explanation. This may be given 8 points if considered to be a highly attractive feature, and the machine that incorporates it is awarded 6 benchmark points. The others, which do not have pivoted injection units, are given none. If this feature is rated lower than other criteria and given only 2 points, the machine with a pivoted injection unit is still given 6 benchmark points and the others none. The overall effect is that, in the first instance, a pivoted injection unit adds 48 points to the total for the machine concerned and 12 in the second.

▲ Replacement period, depreciation and residual value.

There are differences between large and small companies in how replacement and depreciation are handled. In a large organisation with a set policy concerning asset value depreciation, deciding this and associated questions about residual values may not be within the responsibility of the technical executive in charge of choosing plant and equipment and running the moulding shop, even if he is at director level. He simply has to follow the policy laid down from above.

It is usual for 10 years to be taken for depreciation and the first cost of the machine is written off over this time in equal annual amounts. In effect, the depreciation period also becomes the replacement period because company policy will forbid the disposal of the machine until it has been fully amortised. Most large companies are run on the basis of budgetary control, by which each department has an annual amount of money to spend on capital equipment, and this normally allows for the replacement of written off assets. If allowed budgeted funds for machine replacement, the executive concerned will scarcely refuse to spend them and a machine will be replaced just as soon as it passes the end of its depreciation period. The replacement period is therefore taken to be the same as the depreciation period.

In a smaller company, the executive in charge of moulding operations - possibly the owner or a major shareholder - may, in contrast to the situation in a larger concern, have both replacement and depreciation questions within his province and wish to consider them at the time of choosing a new machine. The following remarks are to guide anyone in this position and deal with factors particular to injection moulding machines which should be borne in mind when standard works cost accountancy methods are applied.

The questions one may have in mind will probably include the following: -

(a) What depreciation period should be taken?
(b) Should this be longer or shorter according to the price of the machine?
(c) How should residual value be taken into account?
(d) How should the likely higher cost of a replacement machine be anticipated?

Taking these in turn: -

(a) As stated above, most large companies take 10 years as the depreciation period for their machines, but among smaller companies, whilst 10 years is adopted by the majority, a significant number depreciate over only seven or even five years. Where this is the case, it is part of a deliberate policy to maintain an up to date complement of machines with the object of keeping ahead of competitors.

(b) An expensive machine may be depreciated over 12 or even 15 years but there is not necessarily any point in doing so. The main reason for buying an expensive machine in the first place is not for longer life expectancy but to achieve a high hourly output with a low reject rate and thus achieve a low cost per part produced. Over 10 years, an expensive machine may perform 50 per cent or more cycles than a less expensive machine because of its higher speed. It is probably better and certainly easier to apply the same depreciation period to all one's machines irrespective of first cost and not allot an extended period to the more expensive types.

(c) There are two ways of taking residual value into account. It is either anticipated or dealt with only when it is realised at the time of the final disposal of the machine. Taking the first, the residual value at the end of the depreciation period is estimated and deducted from the sum to be amortised. With the second, the full sum is written off and the actual residual value realised at the end of the life of the machine entered as a credit in the company accounts. This is the more popular method and the more straightforward of the two as the residual value does not have to be estimated.

Inexpensive machines have no or little more value than as scrap at the end of a 10 year depreciation period, so that this question hardly applies to them. It is relevant only to more expensive machines because these retain a greater percentage of their initial value. Because of the effect of continuing inflation, the residual value of an expensive machine in good condition at the end of 10 years may equal 50 per cent of its first cost. Only half of the cost of the machine is therefore written off if the chosen basis for calculating depreciation is to deduct the estimated residual value and the effect of inflation (a further imponderable) is also reckoned with.

Residual value of any machine depends a great deal upon its condition. Even a fairly expensive machine may have little more than scrap value if it has evidently been harshly used and poorly maintained. Oil leaks and excessively worn platen bolt holes are signs of hard use and lack of care. Often, too, when an old machine is taken out of service, a number of vital parts such as nozzle shut off mechanisms, instruments, heater bands and electrical switches may be covertly taken off for spares by maintenance staff and it will be made unsaleable as a working unit.

Overall, however, the depreciation period to choose and how to deal with residual value are not matters of major importance, although many users ponder over them at undue length. It is better to keep the replacement of a machine under constant review, irrespective of over how long a period it is being depreciated in the company's books. Replacement should be considered at least annually after the first few years of ownership and the moment for it will arrive when the amount of money that needs to be invested in a new model is justifiable, taking its higher performance into account, and it is unprofitable to continue to operate the existing machine.

When injection moulding machine development was not as rapid as it became with the introduction of electronic controls in the 1970s, there was little risk in applying a 10 year period both for depreciation and replacement and the two came to be regarded as synonymous. The accelerated development of machines with the introduction of electronic controls (and greatly improved hydraulics) changed matters radically.

There have been many examples, chiefly in the faster moving 'high tech' sectors of the industry, of where advancing technology quickly made machines effectively obsolete and unprofitable well within 10 years. Cycle times for certain components dropped by a factor of four or five within only a few years and production economics became drastically altered.

The 10 year depreciation period has survived as a kind of convention when in many instances it should have been discarded in favour of a policy of constant review of replacement. If it should be decided to replace a machine before it is fully depreciated, the shortfall of the accrued depreciation allotments against first cost is taken into account when reckoning the benefits of the fresh investment, in other words to add it to the cost of that investment.

It also remains true that in many cases a machine can be profitably retained for longer than an arbitrary 10 years. The danger here is that operating costs might be underrated because the machine no longer carries an annual depreciation allotment. There is an argument, for this reason, to continue to load written off machines with a notional annual depreciation allotment even after they have been fully amortised. Holding on to old machines for too long is a trap into which small companies are more likely to fall than larger organisations with fixed depreciation periods and budgetary control. There will be no reluctance by an executive in a large concern to spend a budgeted allocation to replace a fully depreciated machine, even if it has some useful life left in it.

In a smaller company, a fine judgment often has to be made about replacing written off machines, especially when they are still performing well, but this should ignore the fact that there are no more annual depreciation allotments. The main question is whether the ongoing factory overheads and other fixed costs are being spent to advantage. The dominant factor in injection moulding is material costs and machines which make the most economical use of material are the most profitable to employ. These are not necessarily old machines which no longer incur depreciation costs because they have been fully amortised.

Replacement is often needlessly protracted by the disposal of an existing machine. A customer will often say to a supplier that he will buy a new machine from him if he can find a buyer for the old one he wishes to replace. This postpones the conclusion of a deal. The sales forces of some suppliers spend a great deal of their time and effort not so much selling their own wares as trying to find new owners for old machines that prospective customers wish to dispose of. Sometimes, too, several salesmen of a number of different suppliers find themselves trying to sell the same old machines at the same time, which compounds the unsatisfactory nature of these circumstances.

Suppliers generally regard old machines as a drug on the market, even if they are in reasonable working condition, and in many cases will build in an appropriate margin into their quotations for new machines. They take the traded in machine for a certain 'consideration' but in fact dispose of it at a loss as scrap. This loss is taken into account in the price quoted and is therefore not real.

A prospective buyer is probably wise to sell his old machines for scrap himself and tell the supplier to quote on the basis that he is not expected to take a trade in. The overall benefit to the buyer, and one which is commonly overlooked, is simply that the new machine with good profit potential is brought into his plant sooner than would be the case than if a new owner has to be found. One again comes back to making best use of material. An old machine awaiting disposal does not make profitable use of material and is best got rid of as quickly as possible.

(d) Machine prices increased by approximately 50 per cent over the 10 years preceding 1995, the year of publication of this book. A 500 kN machine of a type which cost about £45,000 in 1995 was £30,000 in 1985. Anyone who fully amortised a £30,000 machine over 10 years ended in 1995 would therefore find himself in theory short of £15,000 to replace it. If it is decided to take increasing machine prices into account in depreciating a machine over 10 years, these facts indicate that the first cost should be increased by 50 per cent. Nothing is lost if continuous inflation, even if at a low rate, and the fall of the £ Sterling against harder Continental currencies (until such time as there is a Single European Currency) is assumed.

In this same vein, there is an argument, if one is considering the purchase of a machine in a hard Continental currency, to buy the money forward at the time of placing the order, unless the supplier agrees to hold the price prevailing at the time of the order.

glossary

Glossary of injection moulding terms

It may be valuable for a reader to understand certain technical terms employed in injection moulding as he may encounter them in the course of selecting a machine and wonder not only what they mean but whether they are in fact relevant. The following list is not exhaustive.

▲ Materials terms

■ **Polymer**
■ **Plastics**
■ **Resin**

The word 'polymer' is loosely used to mean the same as 'plastics', as in: 'Moulder X processes 500 tonnes of polymer a year.' Strictly it means the main ingredient of a material of a certain kind. The polymer is the base ingredient produced by a chemical refining process but is mixed, or 'compounded', with various additives. These may be colour, lubricant to make it easier to process, and others to improve its gloss, flexibility or resistance to ultra violet light. Fire retardant additives may also be included. The resultant mixture is normally referred to as the 'material'. 'Resin' is an alternative word for material but becoming out of date. It was coined in the early days of plastics as the new synthetic materials were likened to the tree product.
Note that 'plastics' almost always takes a final 's'. No one in the plastics industry talks about 'a plastic cup', for example, always 'a plastics cup'. Plastic, without an s, is an adjective used to mean that the material can be moulded in its molten state, although the full term is 'thermoplastic'. Thermo means heat. When the material is heated, it becomes plastic and can be moulded.

■ **Compact**
■ **Expanded**

It became necessary to coin the word compact to differentiate plastics mouldings in their normal form from expanded. Most mouldings are in compact material, that is solid all the way through. Expanded materials contain a foaming or frothing agent which is activated by the heat of the process and creates a microcellular or foamed lightweight structure. It is possible to produce mouldings which are compact on the outside but expanded inside.

■ MFI (Melt Flow Index)
■ IV (Intrinsic Viscosity)

The Melt Flow Index is a scale for the flowability of a material in its molten state and approximately correlates to its viscosity. A high MFI material will have a low viscosity and vice versa. A mould with a spiral groove may be used to check the MFI of a material. The maximum length of spiral that can be moulded under given conditions is an indication of the MFI. It is advisable to check the MFI of successive batches of material being processed against the set of moulding conditions predetermined for it - stroke distances, pressures, temperatures and so on - as any change will affect the quality of the parts produced. Intrinsic Viscosity is a scale used for the flowability of thermoplastic PET (polyester) materials.

■ Anisotropy

This is the opposite of isotropy, the normal relaxed state of the molecular polymer chains in a plastics melt when not subject to pressure. When forced under high pressure into a mould, their arrangement and alignment become different from the isotropic state and, because they are quickly cooled, the part solidifies before they can revert to this. An anisotropic molecular chain arrangement and alignment is thus frozen into the moulded part and this affects its strength and how it shrinks. Shrinkage is less in the direction of the alignment of the chains than across, and the part will break more easily in line with the chains than across them.

These factors have to be taken into account in the mould design. The anisotropy of a moulding depends to a large extent upon its gating, or where the material enters the mould impression. If the strength of a part is critically impaired by its anisotropy and the problem cannot be overcome by changing the mould design, it may be annealed, that is placed in an oven for a time during which the molecular chains will revert to their isotropic, or stressless, state.

■ Newtonian, non Newtonian

Terms mainly used in rheology, the study of liquid flow. Liquids like water flow in an easily predictable manner and are said to be 'Newtonian' in behaviour so that calculation of their rate of flow is a straightforward. On the other hand the viscosity of a plastics melt decreases the faster the rate of flow because the molecular polymer chains elongate and slide over each other more easily. This is called 'non Newtonian' behaviour and is more difficult to predict, although there are applicable mathematical formulae. Some materials, including PE, tend to behave in a non-Newtonian manner to a greater extent than others.

■ Sinking, sink marks

Chiefly liable to occur with PP, sinking is shrinkage of the material which takes place in the mould as the moulding cools to the extent that, in certain areas, the surface of the moulding is pulled away from the face of the mould. A slight depression, known as a sink mark, is formed.

▲ Mould terms

■ Impression, cavity, core

In the main text of this book, the terms 'impression' and 'cavity' have been used to mean the same and such usage is common in the injection moulding industry. Strictly speaking, however, the terms are distinct. The impression is the space in a mould formed when the two halves come together. One half contains the cavity and the other the core. The cavity half is usually that mounted on the fixed platen, and the half with the core on the moving platen. The cavity is the female half and the core the male. The impression is filled with plastics material. The precise terms, as explained here, are used to define the parts of the mould, for example the cavity plate and the core plate.
The word 'core', as well as meaning the male part of the mould, may also apply to a secondary device used to form a hole or depression in the moulding at an angle to the parting line. A device of this kind is known as a 'core puller' or 'core pulling attachment'.

■ Bolster

A mould is made up of a number of plates, each of which has a specific purpose, and the collective assembly is known as the bolster.

■ Parting line
■ Parting line injection
■ Off centre injection
■ Line of draw
■ Core pulling

A fundamental question which determines the design of a mould is where the parting, or 'split', line should be. This is the line around the part to be moulded which divides it such that one side will be in the fixed half of the mould and the other in the moving half. If a beaker is taken as an example, the parting line will be around the rim. This makes it possible to mould the beaker easily because it can be withdrawn from the female half of the mould in a straight line when the mould is opened. It is subsequently ejected off the male core of the moving mould half. (This is the usual method but there are examples of the reverse procedure.)

A beaker is a simple example but there may be a number of alternatives for the parting line with mouldings of more complex shape. These have to carefully considered so that the best way of making the mould can be decided. The parting line of most moulds is in one plane and the two halves of the mould have flat mating faces, but for some complex shapes the parting line may be curved or in two flat planes meeting at an angle of, say, 120° and the mating faces of the mould halves will be shaped accordingly. The mating, or meeting, faces of the mould are the surfaces which match up accurately when the mould closes and are clamped together.

In most instances material is injected into a mould through a hole in the centre of the flat face of the fixed half. This is known as centre line injection, but injection may be done from the side of the mould at the parting line. A channel from the outside of the mould leading to the cavity is provided. This method, known as 'parting line injection', has practical advantages for certain kinds of work and machines built for it have a variety of arrangements for the disposition of the injection unit. They are popular for insert moulding, by which metal parts such as threaded nuts or 'bosses' are first inserted in the desired places in a mould. When the plastics is injected, the inserts are firmly secured in the moulding. Some insert moulding machines have a shuttle table or a turntable to allow loading the inserts to be done in one mould while another is going through the cooling stage.

There is a variation of centre line injection known as off centre injection. The injection unit is not on the dead centre line of the machine but parallel to it a short distance away, either horizontally or vertically according to type of machine. It may be advantageous to mould a long article by injecting not into its centre but at one end, and off centre injection permits this. This is also known as 'eccentric' injection.

The line of draw is the direction in which the moving platen travels when opening the mould and is along the horizontal axis of the machine. Taking the example of a beaker, the line of draw is at a right angle to the plane of the rim. Other mouldings of a more complex shape may have depressions in their forms which are not in the line of draw but an at angle to it. An example is a T piece junction for a drainpipe. The long through channel can be arranged in the line of draw so that the moulding can be withdrawn from the fixed mould half in the same manner as the beaker. However, the short piece of tube at 90 degrees to the main part has to be formed around what is known as a core. This is a moving part of the mould which is placed in position when the mould is closed but before material is injected. The melt flows around the core and forms the side branch tube to the main channel.

It is then necessary to take the core out in order that the main channel can be withdrawn from the mould in the line of draw. Placing the core in position before moulding and withdrawing it afterwards is the function of an additional piece of hydraulic equipment known as a core puller, or core pulling attachment.

It is not usual for a machine manufacturer to provide a core pulling attachment as standard on a machine but he may make provision for one in the machine control system so that the attachment can be fitted if ever needed. This feature provides for the correct sequence of movements by the core puller and ensures that the core is in position before moulding and pulled out again before the mould opens. This is known as 'interlocking', that is one movement cannot be made until a necessary previous movement has been performed. A positional sensor signals that the core puller has safely retracted and the mould cannot open until this signal has been given. Some complex moulds, such as those for bottle crates, may need as many as a dozen core pullers. A trade moulder may be advised to have provision for two core pullers on each of his machines.

It is possible to fix a core pulling attachment to the moving platen so that it travels with the moving half of the mould. This makes it possible for the core to be extracted while the platen is moving rather than having to wait for this to be done before opening the mould. Cycle time may thus be reduced. Core pulling 'on the fly' is how this is described but the machine control and hydraulic system must be suitable for it and permit what are called 'parallel' movements. This is an expression from computer terminology and is distinct from 'serial'. If a machine can perform only serial movements, each movement has to be done at a time individually.

■ Family moulds
■ Mould suites

A family mould has a number of different impressions to produce several related parts at once.The top and bottom of a box, for example, may both be made together in a family mould. A mould suite is a collection of tools each of which is to make a different part for a certain product, for example a domestic electrical appliance such as a vacuum cleaner. A new model of vacuum cleaner will be said to need a complete new suite of moulds.

■ Book moulds

Most moulds are made in two halves which can be separated, but a book mould is vertical and its top is hinged and opens like a book. Book moulds are run on machines with vertical clamping units and have certain advantages for insert moulding and footwear production. The machines may have a shuttle table or a turntable on which a number of book moulds can be arranged.

■ Stack moulds

A stack mould is like a double decker sandwich and has two openings, one on each side of a central plate. There are cavities in each opening and in effect twice as many mouldings can be made at the same time as with a conventional mould with one opening. Stack moulds are used for high volume production of parts such as margarine tubs.

■ AISI

AISI (American Iron & Steel Institute) designations for different kinds of steel are internationally recognised and the letter 'P' is used for steels for general purpose injection moulds. For example, P20 is a widely used grade with good machining properties which may be hardened for long wear.

▲ Machine and processing terms

■ Mould open
■ Mould close
■ Cycle time

The part of the machine cycle during which the moving platen retracts and stays open for long enough for the moulding to be ejected is known as the mould open phase. It refers to the time that the platen is moving backwards plus the time it is stationary in its fully open position.
Mould close is the phase during which it moves forward from its retracted position to close the mould plus the time it is stationary in the forward position. Injection and cooling take place once the mould is closed. The moment that the moving platen begins the mould close phase is taken as the start of a cycle. Cycle time is measured from the moment that the platen starts to move forward on one cycle to the same moment on the next.

■ Screw rotate

The phase during which the screw is rotated to plasticise material for the next shot. This term is an example of use of the verb rather than the noun, which tends to confuse non technical people. 'Screw rotation' should be preferred but matters have gone too far now to be changed. This phase is also known as 'recovery' or 'refill'.

■ Melt cushion monitoring

When a screw moves forward on injection it does not come to rest at the extreme limit of its stroke directly against the end cap, the end piece of the barrel. This would cause damage. A margin is left ahead of it and the molten material which remains in the barrel in front of the screw is known as the 'melt cushion'. If there are variations in melt viscosity, the screw may tend to make a longer or shorter stroke on some cycles than on others and the melt cushion will thus be smaller or greater in volume. 'Melt pad' is an American term for melt cushion.
A method of monitoring machine performance is to compare the distance that the screw travels on each injection stroke against what it should be under ideal melt viscosity conditions. This is known as 'melt cushion monitoring'. It is one of a number of means available for the purpose of maintaining consistent

moulding quality and provision for it is normally standard on a high grade machine. If the melt cushion varies considerably over a number of cycles, it is an indication of something being amiss.
A melt cushion monitoring system may have means to restore the desired cushion size if an exceptionally long stroke is made.

■ Accumulator

A device which supplies hydraulic oil at high pressure to assist the machine pump when this is required. It is an alternative to fitting a more powerful pump to the machine and is chiefly used to provide a boost for injection when this has to be done at high speed and pressure, as in thinwall packaging moulding. Some machines use accumulators to assist the pump for fast operation of the clamping unit as well as injection, and one make of machine runs all functions off a bank of accumulators and none directly off the pump. The pump runs at a steady continuous rate and charges the accumulators. An advantage is that there are no 'spikes' or high peaks of draw from the electricity supply.
Nearly all accumulators used on injection moulding machines are of the bladder type. A bladder made of tough, flexible material is contained in a strong steel cylinder and filled with nitrogen. Oil is pumped into the cylinder around the bladder, which contracts as more oil is pumped in. This compresses the nitrogen. When a flow of high pressure oil is required, the outlet valve of the accumulator is opened. The bladder then expands and forces the oil out through the valve at high pressure into the machine hydraulic circuit. There are also piston type accumulators which have certain advantages over the bladder type but are more expensive. The nitrogen is contained in a cylinder and acts against a piston to pressurise the oil.
Large machines of 10,000 kN or more also usually feature banks of accumulators as extremely big pumps would otherwise be required. A hydraulic accumulator is not to be confused with a melt accumulator, as used on some injection moulding machines and large blow moulding machines. This type of accumulator is a cylinder which is filled with melt. The material is then injected into the mould by a piston.

■ Press

An injection moulding machine may be referred to as a 'press' because its clamping unit superficially resembles a press of the kind used in metalforming, as in a 'sheet metal press'. The actions of the two, however, are distinct and the term is in any case becoming archaic, 'machine' being better.

■ SPC (Statistical Process Control)

This is a technique which was introduced in the 1980s and has its origins in product liability legislation in the US. Broadly this legislation gives consumers right of redress against a manufacturer if a product is faulty and failure causes loss or damage. As a result of its implementation it became necessary for manufacturers to introduce methods to enable sub contract sources from which they obtained materials and parts to be traced. If a manufacturer were to be blamed for a failure of his product, one of his suppliers could be pinpointed as the real culprit if the failure lay in bought out parts or materials. This is known as 'traceability', the ability to trace the origin of any part or material used through to its source.

Manufacturers and their suppliers then had to devise methods to make traceability possible and enable part quality to be guaranteed. SPC thus came into being. The underlying principle, so far as injection moulding is concerned, is that it would be prohibitively expensive and time consuming to inspect each and every moulding produced, but that if the moulding process can be closely monitored it is possible to detect any variations from normal in the behaviour of the machine. For the purposes of SPC these are assumed to cause sub standard mouldings to be produced but they can be identified and rejected.

Application of SPC assumes that material which has been certificated by the materials manufacturer is used. The control system of the machine, if suitably equipped, enables an SPC system to be run and a number of critical machine settings, or 'parameters', are selected to be monitored. Taking one of these, overall cycle time, for purposes of explanation, this may be 6 seconds for the moulding concerned. An acceptable tolerance above and below 6 seconds is determined and an Upper Selected Limit (USL) and a Lower Selected Limit (LSL) are set in the control system. If an overall cycle time falls outside either of these limits, the part concerned can be rejected by a diverter mechanism. This directs it to one side after it has been ejected from the mould.

At the end of the production run, an SPC certificate is printed out from the machine and a copy is sent with the parts to the customer as a reassurance that they have been made under SPC conditions. The moulder and the customer retain copies of the certificates against any future possibility that a part needs to be traced back to its source.

SPC has become widely practised in the motor industry and moulders supplying this sector because of the strong US influence in this field. Medical and certain other industries in which product quality is vital have independently evolved their own methods of ensuring impeccability along slightly different lines. SPC should not be confused with SQC (Statistical Quality Control) which involves inspection of parts produced as distinct from monitoring the process by which they are made.

■ Thermoset and elastomer moulding
■ LSR (Liquid Silicone Rubber), LIM (Liquid Injection Moulding)

This book has been written on the assumption that the reader is solely interested in moulding thermoplastic materials, but there are others which can be injection moulded but behave differently under the application of heat and require a slightly different type of machine. Nearly all manufacturers supply machines to mould thermosets and elastomers but some make more of a speciality of these than others and have gained a distinctive reputation in this field. There are a few manufacturers of machines to process LSR.

Thermoset materials are lower in price than thermoplastic and include phenolic, urea, epoxy and melamine. Cycle times are long because the materials need to be heated up to a relatively high temperature in the mould rather than cooled down. This reduces hourly throughput rates so that the price advantage is offset. Mouldings are made in thermoset materials where their specific advantages apply, for example dimensional stability and resistance to high temperatures and electrical tracking (current travelling over a moulded surface).

Thermosets were the original plastics moulding materials but with the increasing improvement and use of thermoplastics now represent only a minor proportion of all kinds used in injection moulding. Their essential characteristic is that heat is applied to them so that they become hard. This is called 'crosslinking', meaning that the molecules all link up with each other end to end and side to side. Once this has happened, the moulding cannot melt and stays rigid under the application of heat. Thermoset saucepan handles exemplify the application of these materials where resistance to high temperature is necessary.

Thermoset materials such as phenolic are in powder form in their raw state, although some granular grades are available. Thermoset polyester is a bulk wet dough supplied in canisters. Machines for this material have special large hopper cylinders from which it is forced through into the injection unit.This material is known as BMC (Bulk Moulding Compound) or DMC (Dough Moulding Compound).

The preliminary heating stage in the barrel of a thermoset machine is up to only a relatively low temperature sufficient to liquidise the material so that it can be injected into the mould. In liquid form the viscosity is extremely low and under pressure the material will tend to run out of the cavities and between the faces of the mould even although these are clamped tightly together. Thin randomly shaped wafers called 'flash' are formed around the mouldings at the parting line of the mould and need to be removed in deflashing equipment, although modern machines with fine control of the process can eliminate or minimise flash formation. Need for deflashing is a major drawback of thermoset moulding and a further reason for the growing use of thermoplastic materials in place of thermosets.

Elastomers are flexible materials based on natural or synthetic rubbers and, like thermosets, are heated in the mould. This vulcanises, or cures, the material, after which it is tougher and more durable than in its uncured state. These materials are usually supplied as coils of uncured strip about 35 mm wide. The strip is unwound from the coil and into the injection unit as the screw rotates.

LSR moulding involves drawing two different liquids from separate storage cylinders into the injection unit and mixing them together, rather as a two component adhesive is prepared. Typical mouldings made in LSR include flexible medical tubing components which in use need to be sterilised by heat.

Processing these materials is mostly performed on purpose-built machines but it is possible to have dual purpose machines which, by interchanging the screw/barrel assembly and making other adjustments, can be used for either thermoplastics or thermoset and elastomer moulding. LSR moulding is highly specialised and interchangeability is less practical. Moulding in LSR is sometimes referred to as LIM (Liquid Injection Moulding).

■ DST (Direct Screw Transfer)

Thermoset injection moulding machines are sometimes called DST machines, meaning that the material is directly transferred from the screw into the mould. The term was coined when this method of moulding thermosets was introduced and supplanted an older technique simply known as 'transfer' moulding. By this method the screw produced a slug of semi rigid material which was injected under pressure into the mould by a second stage plunger system. DST was an improvement as it did away with the mechanical complexity involved in transfer moulding. Transfer moulding is still applied with compression moulding presses to make certain products.

■ FIFO

This is simply an abbreviation for 'first in, first out' and is chiefly applied to the design of the injection units of rubber injection moulding machines. Rubber in the injection unit is in its uncured state but could undesirably cure if it stayed there for too long and absorbed too much heat. The design therefore avoids this by ensuring that, on injection, the first material to exit the injection unit is the first that entered it. None therefore stays behind and soaks up more heat than it should.

■ Release agent

The surface of a mould cavity may be sprayed from an aerosol can with a proprietary liquid, a release agent, to prevent mouldings from sticking to it and making ejection difficult. Release agent is mainly used in rubber and thermoset moulding. The materials are of extremely low viscosity in their molten state and, without the application of release agent, tend to penetrate the microstructure of the mould surface so that they do not come away easily.

■ Intrusion moulding

A method of making thick walled mouldings. Material is not injected in the usual way by the screw making a forward movement. The screw is rotated so that it acts as a pump and material is propelled out through the nozzle and into the mould. Pressure applied to the melt is not as high as by the normal method but is sufficient for a thick walled moulding to be produced. Heavy mouldings can be made as the screw can be rotated for as long as required to fill the mould.This is also known as 'flow moulding'.

■ Injection compression moulding

A method of making thinwalled parts by squeezing the melt. The mould is not fully closed when an amount of material is injected. It is then closed under high pressure so that the material is compressed to a thin wall thickness. The technique is also known as 'injection stamping'.

■ Metal injection moulding
■ Ceramic injection moulding

These processes involve mixing powdered metal or ceramic materials with polyethylene, which acts as a carrier so that mouldings made of the mixture can be produced on an injection moulding machine in the normal way. The mouldings are then placed in an oven so that the PE carrier is vapourised off, leaving compacted finished parts. Components are produced to high dimensional accuracy but these processes are extremely specialised and belong more to metal and ceramic production technology than plastics injection moulding.

■ Injection blow moulding
■ Preform moulding

Injection blow moulding is a method of making bottles. A small tube is first injection moulded. The part below the neck is then inflated to the desired shape while the material is still hot and in a semi molten state. Most bottles are, however, made by extrusion blow moulding on another type of machine. A tube is continuously extruded and cut off into lengths, or 'parisons', and these are inflated. Injection blow moulding is used to make small bottles as larger bottles can be made more cheaply by extrusion blow moulding.
Preform moulding usually involves injection moulding preforms in PET thermoplastic polyester material which are transported to the premises of a bottler and blow moulded adjacent to the liquid filling lines. This is therefore a two stage process involving intermediate transport over a long distance. A preform, which looks like a test tube, is reheated (softened) and blown to shape in what is consequently known as a reheat blow moulding machine.

appendix

Cavity pressure and clamping force calculation

The following formula may be used for the calculation of clamping force but, as will be explained, there are adjustments to be made to the figure obtained. As a further guide, a number of typical mouldings are illustrated on page 24 and the sizes of machine on which they are produced are given.

$$\text{Clamping force kN} = \frac{\text{Projected area cm}^2 \text{ x cavity pressure bar}}{100}$$

Projected area can be calculated from the mould drawing, a sample of the moulding or a model. In simple terms, it is the area of the surface of the moulding as if one were looking at it from the centre of the moving platen towards the injection unit. A figure for cavity pressure is obtainable from the graphs on the following page or by reference to a technical manual.

In order to use the graphs it is necessary to know the maximum flow path length and the minimum wall section. The term 'flow path' means the route taken by the material as it emerges from the 'gate', the entry point into the mould cavity, and fills it. The length of the flow path is the distance from the gate to the farthest part of the cavity and it should be possible either to refer to the mould drawing for this dimension or measure the moulding itself. The flow path may not be in a straight line from the gate but involve a number of corners and turns in the case of mouldings of complex shape.

Wall thickness means the thickness of the individual sections and any stiffening ribs of the moulding. The wall thickness of a disposable vending cup may be 0.5 mm and that of a bucket or washing up bowl 3 mm. Wall thickness usually varies in different parts of a complex moulding such as a TV cabinet or a car bumper, but it is the thinnest that have to be taken into account for the calculation.

Material usually runs through various narrow channels when it is filling the mould and the narrower that these are, the greater the cavity pressure required. The width of the channels in the mould corresponds to the wall thicknesses of the moulded part. Melt does not run freely through a channel in a mould but chills and solidifies against the sides. Solidified material therefore gradually builds up and increasingly constricts the flow of melt through the middle of the channel. In practice there is a limit to the length of a flow path through a channel that can be moulded because of this effect.

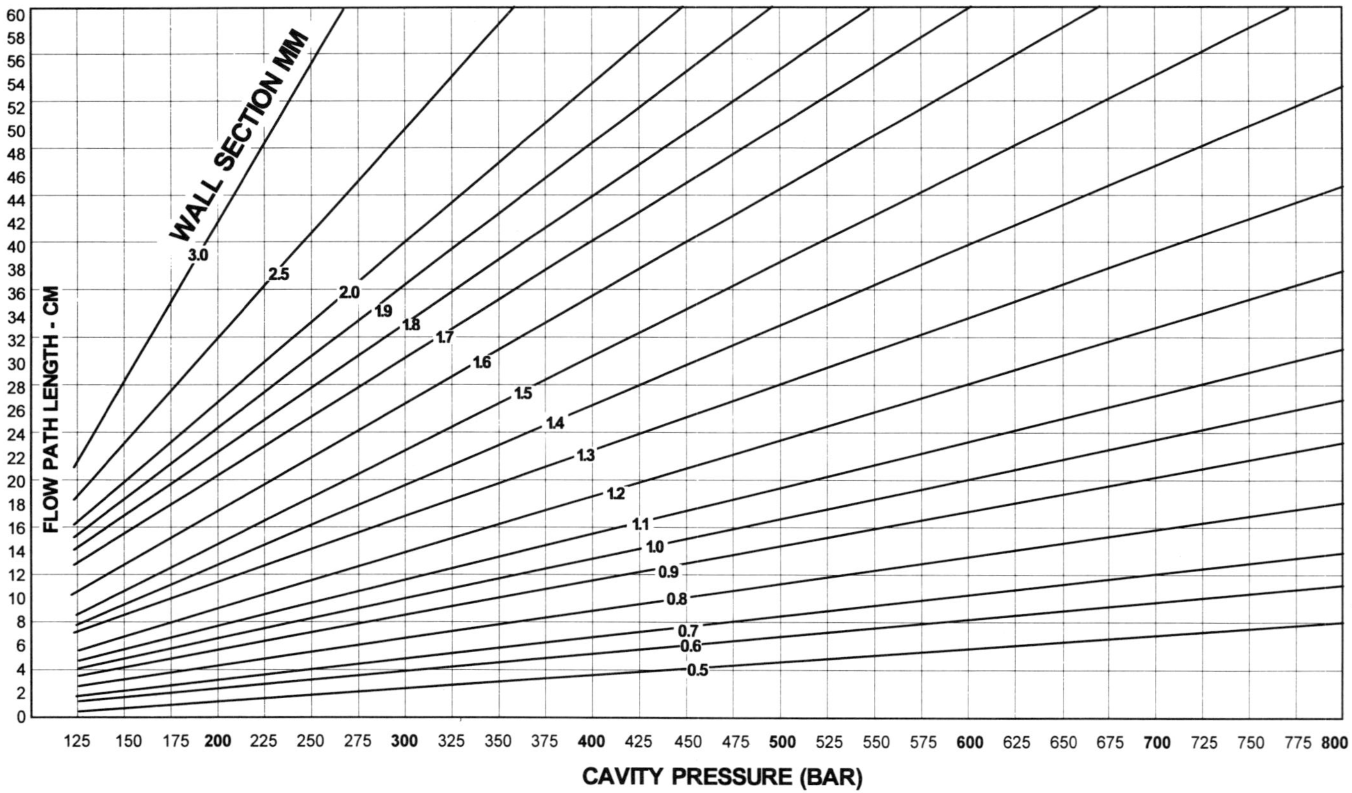

Cavity pressure graphs: To find the cavity pressure for a moulding, read off the flow path length in cm from the vertical axis and down to the horizontal axis from the intersection with the relevant inclined graph for wall thickness. The figure obtained may need to be adjusted according to type of material (see text). There are practical limits to flow path length according to type of material. (Reproduced with acknowledgements to Billion UK.)

The practical limit for any material is stated in the maker's data. To find cavity pressure from the graphs, take the flow path length on the vertical scale and go across horizontally to the appropriate inclined line for the minimum wall thickness of the moulding. There is a series of lines for various wall thicknesses from 0.5 to 3 mm. Mark the point where the horizontal line intersects the wall thickness graph and then drop down from this point to the lower horizontal scale. The cavity pressure is shown where this line intersects it.
A number of graphs of this kind are available in various works of reference, also tables of figures, but precisely identical results are not necessarily forthcoming if one uses data from different sources. Each reference is based upon an individual set of experimental data and there could be differences between these. Most mouldings are made with a cavity pressure of between 200 and 300 bar, or perhaps up to 600 bar, anything more being exceptional and found only in respect of extremely thinwall parts.A number of variables not mentioned here also affect the actual clamping force needed and some of these are not easily quantifiable. Rate of injection is one. If material is injected into a cavity extremely quickly, there is less time for it to cool and effectively constrict the width of the channels, as explained above, and clamping force can be reduced. However, this cannot be done with materials that are liable to scorch under such treatment. Materials of this kind are said to be 'shear sensitive'. Any precalculation of clamping force has to be regarded as an estimate and the actual value finally determined by a practical trial.Having obtained a figure for cavity pressure, it is applied to the formula. How this works out is as follows if it is imagined that the moulding has a projected area of 600 cm^2 and that the cavity pressure is 300 bar:-

$$\text{Clamping force kN} = \frac{\text{Projected area 600 cm}^2 \text{ x cavity pressure 300 bar}}{100} = 1{,}800\text{kN}$$

There is a final adjustment to be made. The type of material has to be taken into account as some are more viscous than others and involve greater cavity pressure. The figure obtained from the formula, 1,800 kN in this instance, is applicable to PS (polystyrene) but needs to be increased for other materials as listed below. The kN figure is therefore multiplied by the factor shown:

Material	Factor
PE (polyethylene)	1 to 1.3
PP (polypropylene)	1 to 1.2
PA (polyamide, nylon)	1.2 to 1.4
ABS (acrylonitrile butadiene styrene)	1.3 to 1.5
PMMA (acrylic)	1.5 to 1.7
PC (polycarbonate)	1.7 to 2.0
PVC	2.0

If, for example, we are going to make our moulding in ABS, 1,800 kN has to be multiplied by from 1.3 to 1.5, and taking 1.5 we arrive at 2,700 kN. This figure can be taken for the purpose of machine selection and checked under the practical conditions of a mould trial.

It has to be appreciated that *cavity* pressure is not the same as *injection* pressure, which is the pressure applied to the melt at the front of the barrel by the screw on the injection stroke. There is a great deal of difference between the two. Cavity pressure is the pressure required by the melt once it has entered the cavity in order to fill it. Injection pressure is the pressure required for the melt to be forced out of the nozzle of the machine and through the channel or channels to the cavities and for there still to be enough left to provide sufficient cavity pressure. Cavity pressure is calculated from the measurements and design of the cavity, and the injection pressure required is taken from the figures given by the manufacturer of the material.

When a multi cavity mould is made, the channels to the cavities farthest from the point where material enters the mould may be gradually widened so that the constrictive effect described above is counteracted. It is desirable at the mould design stage to provide channels of similar length to each of the cavities, but this is not always possible. The sizes of the gates into the farthest cavities may also be enlarged compared with those nearer to the machine nozzle on the material flow path. The ideal is for each cavity to be filled with an equal amount of material and all at the same time. Designing the mould to have equal lengths of flow path to each cavity and suitably adjusting cross sectional sizes of the channels and the gates is known as 'balancing' the cavities.

A fresh calculation of clamping force required should be made if an old machine is to be replaced by a new model. One should not simply buy a new machine with a similar clamping force. Platen sizes relative to clamping force have increased over the years. The original machine may have been bought chiefly on account of its platen size, which needed to be big enough for a particular mould, with clamping force as a secondary consideration. A modern replacement machine with the same clamping force will probably have platens which are larger than necessary (assuming the same mould size). A new machine with adequate platen area but less clamping force will be less expensive than one with the same clamping force as the old machine.

further reading

Key letters given here in bold type indicate the main subject area or areas of the book concerned. Titles do not in all cases completely reflect the scope of the text.

■ **A.** Part design for injection moulding.

■ **B.** Material characteristics and selection for various applications.

■ **C.** Mould design and construction.

■ **D.** Machines and processing control.

'Moulding machines and moulds for plastics processing' by Gianni Bodini and Franco Cacchi Pessani. Negri Bossi Division, John Brown Plastics Machinery, Hollis Lane, Chesterfield, Derbyshire S41 0DS. **BCD.**
Although this book is published by a manufacturer of injection moulding machines, its content is applicable to all makes and types. It includes a description of the behaviour of various plastics materials during processing and related matters such as post moulding shrinkage. Detailed tables of data for the most widely used thermoplastic and thermoset materials are included and the design of machines of various types is described. Graphs are given for defining cavity pressure and clamping force and the data upon which these are based is derived from theory and the authors' practical experience. A copiously illustrated section deals comprehensively with various kinds of mould design and construction features.

'Injection Molding Handbook' 2nd edition by Donald V. Rosato and Dominick V. Rosato. Chapman & Hall, International Thomson Publishing, Berkshire House, 168-173 High Holborn, London WC1V 7AA. **ABCD.**
One of the most comprehensive books on the subject available, this work covers part and mould design as well as material and machine selection. How to calculate the required number of mould impressions for a given rate of output is described. The book also deals with various aspects of processing including quality control and preventive machine maintenance. There are also chapters on equipment for conveying and drying materials, water chilling and welding by ultrasonic and other methods. Systems for plant financial control and product costing are also dealt with.

'Injection moulding of thermoplastics' by Christof Kempf and Michael Renger. Published by Dr Alfred Huethig Verlag and obtainable from Battenfeld UK, 6 The Valley Centre, Gordon Road, High Wycombe, Bucks HP13 6EQ. **BCD.**
This publication is a collection of nine lectures given at the Ludenscheid Plastics Institute, Germany. A description of the molecular behaviour of various plastics materials is given and a guide to determining cavity pressures is contributed by Dipl.Eng Erhard Genster of the Iserlohn Technical College. The interaction of pressure, temperature and speed during the mould filling phase is described and a detailed troubleshooting chart is included.

'Understanding Injection Molding Technology' by Herbert Rees. Carl Hanser Verlag, Postfach 86 04 20, 81631 Munich, Germany. **CD.**
Written in easily understood language, this book introduces the reader to basic technical information needed to understand the injection moulding process. It also deals with compression and transfer moulding as well as extrusion and thermoforming.

'Physics of Plastics', by Arthur W.Birley, Barry Haworth and Jim Batchelor. Carl Hanser Verlag, Kolbergerstr 22, D-8000 Munich 80, Germany. **AB.**
Chiefly aimed at product designers, this book deals with the material properties that determine the processing and performance of plastics components. Reasons are given for failures in service because of brittleness, cracking and other defects, together with advice on gate location to avoid stress in moulded parts.

'Designing with Plastics', by G.W.Ehrenstein and G Erhard. Carl Hanser Verlag, Kolbergerstr 22, D-8000 Munich 80, Germany. **AB.**
A diagram showing a recommended sequence to follow in designing plastics components is included in this book, which also deals with material selection.

'Guide to surface defects on thermoplastic injection moulded parts.' Engel UK, Apollo Way, Tachbrook Park, Warwickshire CV22 6RW. **D.**
A newcomer to injection moulding will have little idea of how to judge moulding quality but this book solves this problem by giving an exhaustive list of possible moulding faults, the causes of which are described. Liberally illustrated, it is based on the results of a project by Dr.Ing. Paul Thienel of the Plastics Institute for Medium-sized Companies in Germany.
The **'Material data for injection moulders'** slide rule available from this company is described and illustrated on page 47. **BD.**

'How to use injection process control equipment. Nissei technical manual 8104'. Obtainable from Trendpam, Unit 24, Barwell Business Park, Chessington, Surrey KT9 2NY. **D.**
A great deal of practical information on machine setting is contained in this booklet, written in elementary terms. Four basic patterns for injection speed and pressure for different kinds of moulding are given and the effects of fast and slow

speeds explained. Examples of difficult mouldings are featured and suitable speed patterns recommended.

'Polymer Products' by D.H. Morton-Jones and John W. Ellis. Chapman & Hall, 11 New Fetter Lane, London EC4P 4EE. **AB.**
This book describes the injection moulding process is easily understandable terms and deals with the selection of materials for various injection moulded products, design factors and processing conditions. Other processes dealt with include RIM (Reaction Injection Moulding) and blow moulding.

'Plastics Engineering Handbook of the Society of the Plastics Industry' edited by Michael L.Berlins. Chapman & Hall, 2-6 Boundary Row, London SE1 8HN. **BC.**
Polymer chemistry and properties of materials, colourants, flame retardants, blowing agents and other additives are dealt with on this book, which also covers mould design and joining and assembly techniques.

'Polymer Processing' by D.H. Morton-Jones. Chapman & Hall, 11 New Fetter Lane, London EC4P 4EE. **BD.**
The processing characteristics of different polymers are fully explained in this book, which describes compounding and other mixing processes and aspects of injection moulding screw design.

'Approaches to automation'. Published by Rapra Technology, Shawbury, Shrewsbury, Shropshire S14 4NR. **D.**
This book of papers presented at a seminar held by Rapra contains practical advice on equipping an injection moulding plant with systems for materials and parts handling and applying modern management practices.

'How to Make Injections Molds' by Georg Menges and Paul Mohren. Sponsored by the Society of Plastics Engineers. Carl Hanser Verlag, Kolbergerstr 22, D-8000 Munich 80, Germany. **C.**
A comprehensive work of reference, this book has chapters on the selection of steel and non ferrous mould materials, and machining by cutting, spark erosion and chemical etching. Procedure for estimating mould costs is described, also a method of determining the optimum economic number of cavities. Runner system design, hot runner systems and the computation of cooling times are also covered.

'Handbook on Thermoplastics Injection Mould Design' by P.S. Cracknell and R.W. Dyson. Chapman & Hall, 2-6 Boundary Row, London SE1 8HN. **AC.**
Written from a practical standpoint and containing essential data on all aspects of the subject, this book may also serve as a general introduction to the injection moulding process. Hot runner systems, provision for cooling and the calculation of part wall thicknesses are dealt with, as are core pulling systems and prototype tooling in non ferrous materials. A formula for the estimation of ejection force

is included.

'Applications of Computer Aided Engineering in Injection Molding' edited by Louis T. Manzione. Carl Hanser Verlag, Kolbergerstr 22, D-8000 Munich 80, Germany. **CD.**
Different applications of computer technology to injection moulding are described in this work. It covers mould fill simulation, gate design, cooling, practical moulding and cost analysis, machine selection and calculating optimum production batch sizes.

'Statistical Process Control. A User's Guide from Dataputer.' Published by Elcometer Instruments, Edge Lane, Droylsden, Manchester M43 6BU. **D.**
This book gives an outline introduction to SPC, explains its purpose and defines the various terms used.

▲ Industry code of practice

'Code of Practice for the design and manufacture of plastics products'. Published by the Institute of Materials, 1 Carlton House Terrace, London SW1Y 5D.
There was no generally agreed way of coordinating the design of plastics parts and producing and testing moulds through to readiness for production until the Institute of Materials, in cooperation with the Gauge and Toolmakers Association and the British Plastics Federation, devised this Code. The process is divided into 12 stages and a team of five people, representing product design, tool design, mouldmaking, injection moulding and material specification, steers the project through according to an action plan. Specimen forms and check lists are given for mould specification, pre delivery inspection and machine trial.

▲ Directories

The following classified directories list UK-based suppliers of all types of plastics processing machinery, including injection moulding machines, as well as ancillary equipment.

'UK Plastics Directory'. EMAP Business Communications, 19 Scarbrook Road, Croydon, Surrey CR9 1QH. Produced annually by the publishers of *Plastics & Rubber Weekly*.

'British Plastics & Rubber Directory'. MCM Publishing, 37 Nelson Road, Caterham, Surrey CR3 5PP. Produced annually by the publishers of the monthly journal *British Plastics & Rubber*.

'Members Directory and Buyers Guide'. Plastics Machinery Distributors Association, PO Box 1414, Dorchester, Dorset DT2 8YH. This publication is compiled every three years to coincide with the Interplas exhibition.

▲ Acknowledgements

Illustrations and diagrams used in this book have generously been made available through the cooperation of the following companies, to whom the author expresses his thanks: -

Billion UK
Engel UK
Hunkar Laboratories
Husky Injection Molding Systems
Mannesmann Demag Hamilton
Negri Bossi Division of John Brown Plastics
Netstal Sales & Services
Sandretto (UK)
Trendpam Engineering.

index

A

B

C

D

M

N

O

P

R

S

T

U

V

W